U.S. Navy
SEAL
PATROL LEADER'S
HANDBOOK

PALADIN PRESS
BOULDER, COLORADO

This book is dedicated to the service men and women of our Armed Forces who have served or continue to serve the ideals of the United States of America. Their devotion to duty enables them to obey the orders of a Commander-in-Chief who refused to serve when he was asked to do so and of Congress, whose majority scorns their service at the same time it controls the appropriations and authority to make and conduct the wars to which these men and women are sent.

U.S. Navy SEAL Patrol Leader's Handbook

Copyright © 1994 by Paladin Press

ISBN 0-87364-778-5
Printed in the United States of America

Published by Paladin Press, a division of
Paladin Enterprises, Inc., P.O. Box 1307,
Boulder, Colorado 80306, USA.
(303) 443-7250

Direct inquiries and/or orders to the above address.

All rights reserved. Except for use in a review, no portion of this book may be reproduced in any form without the express written permission of the publisher.

Neither the author nor the publisher assumes any responsibility for the use or misuse of information contained in this book.

TABLE OF CONTENTS

NAVAL SPECIAL WARFARE FORCES- OVERVIEW

CHAPTER 1   MISSION PLANNING PROCESS

1.0	INTRODUCTION	1
1.1	THE MISSION PLANNING CYCLE	1
1.2	THE MISSION PLANNING PROCESS	1
1.2.1	Receive the Mission Directive	1
1.2.2	Initiate a Security Plan	1
1.2.3	Analyze the Mission	2
1.2.4	Plan the Use of Available Time	3
1.2.5	Submit an Initial EEI Request	3
1.2.6	Formulate an Initial Plan	3
1.2.7	Give Mission Concept	3
1.2.8	Revise Plan, Based On Mission Concept	4
1.2.9	Phase Plan/Diagram the Mission	4
1.2.10	Update EEI Request	4
1.2.11	Submit Support Requirements	4
1.2.12	Issue Warning Order	4
1.2.13	Conduct Preliminary Gear/Personnel Inspections and Rehearsals	4
1.2.14	Update the Plan as Necessary	4
1.2.15	Patrol Leader's Order	4
1.2.16	Briefback	5
1.2.17	Final Inspection, Rehearsals, and Brief	5
1.2.18	Conduct Mission	5
1.2.19	Debrief	5
1.2.20	Submit Post-Op Report	5
1.3	THE PHASE DIAGRAMMING SYSTEM	5

AN OVERVIEW

1.4	ORGANIZATION OF THE PHASE DIAGRAM	6
1.5	MISSION PHASES	6
1.6	PHASE DIAGRAMMING	7
1.6.1	Analyze Each Event	7
1.6.2	Continue the Analysis	8
1.6.3	Prepare Detailed Lists	8
1.7	PLANNING FOR CONTINGENCIES	12
1.7.1	Contingency Checklist	12

CHAPTER 2
ESSENTIAL ELEMENTS OF INFORMATION

2.0	INTRODUCTION	14
2.1	TARGET	14
2.1.1	Enemy Environment	15
2.1.2	Enemy Order of Battle	15
2.1.3	Survival/Evasion/Resistance/Escape (SERE)	15
2.1.4	Miscellaneous	15
2.2	TARGET DEPENDENT EEI	15
2.2.1	Imagery and Graphics	15
2.2.2	Textual Data and Support Materials	15

CHAPTER 3   TARGET ANALYSIS

3.0	INTRODUCTION	16
3.1	TARGET SELECTION	16
3.1.1	Criticality	16
3.1.2	Accessibility	16
3.1.3	Recuperability	17
3.1.4	Vulnerability	17
3.1.5	Effect on Populace	17
3.1.6	Recognizability	17
3.2	TARGET SYSTEMS	17
3.3	MAJOR TARGET SYSTEMS	18
3.3.1	Railway Systems	18
3.3.2	Highway Systems	18

TABLE OF CONTENTS

3.3.3	Waterway Systems	18
3.3.4	Airway Systems	18
3.3.5	Communication Systems	19
3.3.6	Power Systems	19
3.3.7	Water Supply Systems	20
3.3.8	Fuel Supply Systems	20

CHAPTER 4   MISSION CONCEPT

4.0	INTRODUCTION	21
4.1	MISSION CONCEPT (FORMAT)	21
4.2	RULES OF ENGAGEMENT (ROE) CONSIDERATIONS	22

CHAPTER 5
WARNING/PATROL LEADER'S ORDERS

5.0	INTRODUCTION	24
5.1	THE WARNING ORDER	24
5.2	WARNING ORDER (FORMAT)	24
5.3	RECOMMENDED BRIEFING ASSIGNMENTS	26
5.4	PATROL LEADER'S ORDER	26
5.5	PLO SEQUENCE	27
5.5.1	Five Paragraph Sequence	27
5.5.2	Bad PLOs	27
5.6	PRE-PLO CHECKLIST	28
5.7	PRESENTATION	28
5.7.1	Briefing Techniques	28
5.7.2	PLO Introduction	29
5.8	PLO FORMAT	30
5.8.1	Situation	30
5.8.2	Mission	30
5.8.3	Execution	33
5.8.4	Admin and Logistics	35
5.8.5	Command and Signals	36

CHAPTER 6   BRIEFBACK

6.0	INTRODUCTION	39
6.1	BRIEFBACK (FORMAT)	40
6.1.1	Situation/Mission	40
6.1.2	Intelligence	40
6.1.3	General Overview	42
6.1.4	Insertion Method	42
6.1.5	Routes	42
6.1.6	Actions at the Objective	43
6.1.7	Extraction Method	43
6.1.8	Rendezvous/Evasion and Escape Procedures	43
6.1.9	Communications	43
6.1.10	Medical	44
6.1.11	Closing Statements	44
6.2	BRIEFBACK PRESENTATION	44

CHAPTER 7   POST EXERCISE/OPERATION REPORTS AND INTELLIGENCE DEBRIEFS

7.0	INTRODUCTION	47
7.1	POST-EXCHANGE/OPERATION (FORMAT)	47
7.2	INTELLIGENCE DEBRIEF GUIDE	48
7.3	DEBRIEF GUIDE (FORMAT)	49

APPENDICES

APPENDIX A   NSW INTELLIGENCE

A.1	TARGET INDEPENDENT EEI - ENVIRONMENT	56
A.1.1	Obstructions/Constructions	56
A.1.2	Order of Battle	59
A.1.3	SERE	70
A.1.4	Miscellaneous Information	71

TABLE OF CONTENTS

A.2	TARGET DEPENDENT EEI	72
A.2.1	Imagery and Graphics	72
A.2.2	Textual Data/Support Materials	73
A.3	TARGET ANALYSIS CHECKLIST	76
A.3.1	Administrative Data	76
A.3.2	General	76
A.3.3	Specific	79
A.3.4	Conclusions	80

APPENDIX B   VESSEL CHARACTERISTICS AND CAPABILITIES

B.1	SMALL CRAFT OPERATIONS	82
B.1.1	General	82
B.1.2	Mission Planning Considerations	82
B.1.3	Coordination	82
B.1.4	Execution	83
B.2	COMBAT RUBBER RAIDING CRAFT (CRRC)/INFLATABLES	84
B.3	SPECIAL BOAT CHARACTERISTICS	86
B.4	FLEET BOAT CHARACTERISTICS	88
B.5	SHIPS CAPABLE OF TRANSPORTING THE SEAFOX	90
B.6	NAVAL GUNFIRE SUPPORT SHIPS	92

APPENDIX C   AIRCRAFT CHARACTERISTICS

C.1	FIXED WING AIRCRAFT	95
C.2	ROTARY WING AIRCRAFT	97

APPENDIX D   WEAPONS AND DEMOLITIONS

D.1	U.S. SMALL ARMS	100
D.2	SOVIET/WARSAW PACT SMALL ARMS	102
D.3	DEMOLITION CAPABILITIES AND FORMULAS	103

APPENDIX E   COMMUNICATIONS

E.1	COMMUNICATIONS/ELECTRONICS CAPABILITIES	114
E.2	NSW/SHIPBOARD COMMUNICATIONS INTEROPERABILITY	116
E.2.1	NSW/Shipboard SATCOM Interoperability	118
E.3	NSW/E2C INTEROPERABILITY	119
E.4	C3 VAN CAPABILITIES	120

APPENDIX F   BIBLIOGRAPHY AND GLOSSARY

F.1	BIBLIOGRAPHY	122
F.1.1	Amphibious Operations	122
F.1.2	Environmental Areas of Operations	122
F.1.3	Cartography	123
F.1.4	Demolitions	123
F.1.5	Diving	123
F.1.6	Land Warfare	124
F.1.7	Photography	124
F.1.8	Soviet/Eastern Bloc Forces	124
F.1.9	Submarine Operations	125
F.1.10	Targeting	125
F.1.11	Weapons	125
F.2	GLOSSARY	126

AN OVERVIEW

US NAVY SPECIAL WARFARE FORCES

1. INTRODUCTION

The US Navy Special Warfare Command (NAVSPECWARCOM) is the Navy's proponent agency for Special Operations Forces (SOF) and is designated a Major Command (MACOM) on an equal level with the US Army Special Operations Command (USASOC) and Air Force Special Operations Command (AFSOC). Like its brother services' special operations commands, NAVSPECWARCOM is subordinate to the U.S. Special Operations Command (USSOC).

NAVSPECWARCOM is tasked with maritime related special operations in support of Navy and Marine forces as well as the other services. These missions include special mobile operations, unconventional warfare (UW), beach and coastal reconnaissance, counterinsurgency (COIN), special tactical intelligence collection, coastal and river interdiction, and foreign internal defense (FID), i.e. advise, train, assist and/or control friendly forces in the conduct of naval special warfare operations. These missions include such specialized tasks as special reconnaissance (SR), direct action (DA), combat search and rescue (CSAR); recovery, vessel boarding, search and seizure (VBSS), and beach reconnaissance and obstacle clearance in support of amphibious operations. These missions principally focus on coastal and inshore areas, harbor and ports, but also include operations in inland waterways as well as missions executed ashore, usually contiguous with coastal areas. They may, however, be conducted well inland. There are four principal types of units tasked with execution of and/or support of these missions.

US NAVY SPECIAL WARFARE FORCES

2. NAVAL SPECIAL WARFARE UNITS

The key, and best known, Navy SOF element is the Sea-Air-Land Forces - SEALs. The first SEAL teams were formed in 1962 as the Navy's contribution to America's growing counterinsurgency effort. The former underwater demolition teams (UDT) were absorbed into the Seals in 1983. There are seven company-size SEAL teams (commanded by an O-5). These are comprised of ten special operations platoons. The platoons have two officers (O-3 and O-2) and 14 enlisted men divided into two special operations squads with one officer and seven enlisted; the basic planning element for loading into various watercraft. Platoons are identified by phonetic letters within the team and squads by number. For example, 1/ECHO/SEAL 4 (1st Squad, ECHO Platoon, SEAL Team 4). Two or more platoons may be formed into SEAL detachments to accomplish missions requiring more than one platoon and may operate independently of the SEAL team. The SEAL team also has a headquarters platoon with two officers (O-4 and O-3) and six enlisted men. SEAL Team 6, tasked with a counterterrorism and special missions role, is organized somewhat differently. There are also a few small US Navy Reserve (USNR) (also known as Naval Reserve Force - NRF) SEAL elements that habitually train with active force SEAL teams.

SEAL trainees undertake a grueling 27-week Basic Underwater Demolitions/SEAL (BUD/S - "Buds") Course at Coronado, CA, where they become SCUBA qualified. They are also fully trained in reconnaissance, surveillance, target acquisition, small scale raids and direct action missions. SEAL trainees receive training in combat swimming, survival, demolitions, sabotage, small arms, patrolling, individual and small unit tactics, close combat, navigation, first aid, and communications. Selected members are specifically trained as medical corpsmen and radiomen. All are static line parachute qualified at the Army's Airborne Course at Ft. Benning, GA. Once assigned to a unit, they are under six-months probation and receive additional unit specific training. They also undertake more advanced training courses to include various Army Special Forces courses, Army Ranger, freefall parachuting, and sniper. Small arms used by the SEALs include the 5.56mm M16A1 rifle,

AN OVERVIEW

5.56mm M16A1 carbine, 40mm M203 grenade launcher, 7.62mm M60E3 machine gun, different models of 9mm MP5 submachine guns, 12ga M870 shotgun, 9mm M9 pistol, and various types of sniper rifles including the 7.62mm M21 and .50-caliber M82A1 and M500 along with a number of more specialized, "exotic" weapons.

The two SEAL Delivery Vehicle (SDV) teams are responsible for the operation of small submersibles ("mini-subs"). Two types are used. The MK VIII carries two SVD crewmen and four SEAL passengers or cargo and is a wet system. The MK IX carries only an SVD crew of two along with mission equipment. The 20 foot long MK VIII is intended for attachment to SEAL teams for delivery of SEALs to their targets or operational areas. The smaller MK IX is operated by the SDV team to accomplish DA missions either tasked directly to the SDV team or in conjunction with SEAL teams. The MK IX can also carry and launch a MK 32 standoff weapons assembly, a small homing torpedo for attacking distant targets. SDV teams also operate the Dry Deck Shelters (DDS) fitted to nuclear submarines modified as SDV transports or amphibious transports.

There are six *Sturgeon* class SDV troop transports (still capable of operating as attack submarines) in service. The conversions were between 1982 and 1991. Two *Ethan Allen* class submarines were modified as amphibious transports between 1983 and 1988. The life expectancy of the *USS Sam Houston* (SSN 609) and *USS John Marshall* (SSN 611) is into the late 1990s. These replaced the old diesel amphibious warfare submarines, *USS Grayback* and *USS Waho*. The modifications included removal of some ballistic missile launch tubes, conversion of others to diver air locks and storage compartments, and accommodations for 60-plus SEALs. A removable DDS can be fitted to the decks of these submarines and link-up with their hatches. A DDS can carry one SDV plus be used to lockout large numbers of divers while the submarine is still submerged as well as launch and recover inflatable boats. Three SDV troop transports and one amphibious transport are assigned in both the Pacific and Atlantic Fleets.

US NAVY SPECIAL WARFARE FORCES

SDVs can also be transported by converted utility landing craft known as Advanced SEAL Delivery Vessels (ASDV), of which three exit. These are modified 180 foot Landing Craft, Utilities (LCU). An SDV platoon has two officers and 12 enlisted men, qualified not only to operate and maintain the SVDs, but trained in basic SEAL skills. The platoon a will deploy with anywhere from one to three SVDs. An SVD team will have up to five operational SVD platoons and one or two DDS platoons. These platoons usually operate in direct support of SEAL teams. Additional new systems are under development to include the SEAL Tactical Insertion Craft (STIC) and Advanced SEAL Delivery System (ASDS). The ASDS, intended to replace the MK VIII, will be able to deliver eight SEALs ashore using a dry system. The ASDS will have a crew of four. Six of the 40-50 foot mini-subs will be purchased.

Special Boat Squadrons (SBS or SPECBOATRON) are equipped with various types of coastal and riverine small craft. SPECBOATRONs are subdivided into three or four Special Boat Units (SBU or SPECBOATU), designated by two digit numbers. The active SBUs (12 and 20) are further sub-organized into three detachments designated by phonetic letters. Each detachment habitually trains and operates with a specific SEAL team. Besides supporting SEALs, SBSs also provide special boat support to other services. Of the seven SBUs, three are active force and four are USNR. The USNR SBSs perform riverine and coastal patrol and interdiction (CP&I) missions. The SBSs will acquire some new patrol craft in the future to provide a longer range capability. NAVSPECWAR small craft are discussed in Section 4.

The two USNR Helicopter Composite Squadrons (HCS or HELCOMPRON) are each equipped with eight HH-60H Seahawk helicopters (modified Blackhawks). They are tasked with both strike rescue and special warfare support roles. They are capable of search, rescue, infiltration, exfiltration, and resupply missions. SEALs rely heavily on Air Force special operations aircraft for support as well as Marine Aviation helicopters and transports and the Army's 160th Special Operations Aviation Regiment.

AN OVERVIEW

Higher commands, planning and support organizations for these various unite are controlled by Commander NAVSPECWARCOM (O-7) at Naval Amphibious Base (NAB) Coronado, San Diego, CA. Directly under it and co-located with it is the Naval Special Warfare Center (NAVSPECWARCEN) responsible for SEAL selection and training. The Naval Special Warfare Development Group (NAVSPECWARDEVGRU) is located at Fleet Combat Training Center Atlantic (FCTCLANT), Dam Neck, VA. It is responsible for tactics and equipment development.

Two Naval Special Warfare Groups (NSWG or NAVSPECWARGRU) (commanded by an O-6) provide command and control as well as support to the special warfare units stationed on the West and East Coasts. These are roughly equivalent to an Army Special Forces Group. A NSWTG can form task organized Naval Special Warfare Task Units (NSWTU or NAVSPECWAR-TASKU) to control deployed NAVSPECWAR forces. In effect these are small "task forces" comprised of elements drawn from within the NSWG. They operate similar to an Army Special Forces battalion Forward Operations Base (FOB) providing command and control and preparation of elements for mission execution. Additionally, one or more Naval Special Warfare Task Elements (NSWTE or NAVSPECWARTASKELM) may be formed to provide support to smaller detached NAVSPECWAR elements, such as a platoon. Three operate in a fashion similar to an Army Special Forces Advanced Operations Base (AOB).

Four forward deployed "named" Naval Special Warfare Task groups (NSWTG or NAVSPECWARTASKGRU) provide pre-hostility planning and coordination with other services in specific theaters, similar in concept to an Army Special Operations Command (ARSOC). Naval Special Warfare Units (NSWU or NAVSPECWARUNIT), not to be confused with the NSWTU) are theater oriented support staffs similar to theater army Special Operations Support Commands (SOSC). These are principally concerned with coordination and logistical support of deployed NAVSPECWAR forces.

SEAL platoons (reinforced by two additional men) habitually accompany Marine Expeditionary Units (MEU) (reinforced battalion landing team) and Marine Expeditionary Brigades (MEB) (reinforced regiment) on their overseas deployments. Their principal mission in this case is reconnaissance in support of the amphibious force.

3. NAVAL SPECIAL WARFARE ORGANIZATION

Naval Special Warfare Command	Coronado NAB, CA
Naval Special Warfare Center	Coronado NAB, CA
Naval Special Warfare Development Group	Dam Neck, VA

NAVSPECWAR units based on the West Coast and oriented toward the Pacific basin and Southwest Asia are:

Naval Special Warfare Group 1	Coronado NAB, CA
NSWG 1 Detachment 10	
NSWG 1 Detachment 11	
NSWG 1 Detachment 13	
NSWG 1 Detachment Kodiak	Kodiak, AK
SEAL Team 1	Coronado NAB, CA
SEAL Team 3	Coronado NAB, CA
SEAL Team 3 Detachment	Hawaii (?)
SEAL Team 5	Coronado NAB, CA
SEAL Vehicle Delivery Team 1	Coronado NAB, CA
SEAL Vehicle Delivery Team 1 Det Hawaii	Pearl Harbor, HI
Special Boat Squadron 1	Coronado NAB, CA
Special Boat Unit 11 (USNR)	Mare Is, CA
Special Boat Unit 12	Coronado NAB, CA
Special Boat Unit 13 (USNR)	Coronado NAB, CA
Helicopter Composite Squadron 5 (USNR)	NAS Pt. Mugu, CA
Naval Special Warfare Unit 1	Guam

AN OVERVIEW

Naval Special Warfare Task Group
    Seventh Fleet   Pearl Harbor, HI
SEAL Element Western Pacific   Guam
Special Boat Detachment Western Pacific   Guam

NAVSPECWAR units based on the East Coast and oriented toward Latin America, Caribbean, Europe, and Africa are:

Naval Special Warfare Group 2	Little Creek, VA
NSWG 2 Detachment 6	Dam Neck, VA (?)
SEAL Team 2	Little Creek, VA
SEAL Team 4	Little Creek, VA
SEAL Team 8	Little Creek, VA
SEAL Vehicle Delivery Team 2	Little Creek, VA
Special Boat Squadron 2	Little Creek, VA
Special Boat Unit 20	Little Creek, VA
Special Boat Unit 22 (USNR)	New Orleans, LA
Special Boat Unit 23 (USNR)	Little Creek, VA
Special Boat Unit 26	Rodman NAS, PM
Helicopter Composite Squadron 4 (USNR)	Norfolk, VA
Naval Special Warfare Unit 2	Machrihanish, UK
Naval Special Warfare Unit 4	Roosevelt Rds, PR
Naval Special Warfare Unit 8	Rodman NAS, PM
Naval Special Warfare Task Group Atlantic	Norfolk, VA
Naval Special Warfare Task Group Europe	Machrihanish, UK
Naval Special Warfare Task Group South	Rodman NAS, PM

SEAL Team 6 is under the operational control of the Joint Special Operations Command (JSOC) and tasked with counter-terrorism and special missions along with the Army's Delta Force.

US NAVY SPECIAL WARFARE FORCES

Besides major training and support facilities at Coronado NAB, CA, Little Creek, VA and Dam Neck, VA, the Naval Special Warfare Command has developed support facilities at Pearl Harbor, HI and Roosevelt Roads, PR.

SEALs and their supporting units have operated in Vietnam (1962-1972), Grenada (1983), Persian Gulf (1987-88), Panama (1989-90), Gulf War (1990-91) and Somalia (1991-93) as well as many smaller contingency operations. They also participate in major exercises etch as: COBRA GOLD (Thailand), BRIGHT STAR (Egypt), OCEAN VENTURE (Puerto Rico), and FUERZAS UNIDA (Central America). The units are presently being increased in strength and a wide range of new equipment items are under development.

4. NAVAL SPECIAL WARFARE SMALL CRAFT

Since different SPECBOATUs have different primary missions, they are equipped with different types of small craft. SPECBOATUs dedicated to SEAL support and coastal patrol and interdiction use the PB and SWCL. Those dedicated to riverine warfare and SEAL support are equipped with the MATC and PBR. Some units are additionally equipped with high-speed "cigarette boats" confiscated by the Coast Guard from drug runners.

Sea Spectre MK 3 and 4 Patrol Boats (PB). The all-aluminum Sea Spectres were designed as a high-speed weapons platform for Naval Inshore Warfare (NIW) forces. The last MK 3s were acquired in 1977 and the last MK 4s in 1984. They are capable of day and night patrol, surveillance, interdiction, and fire support missions in deep rivers, harbors, coastal and open area environments for up to five days duration. They are equipped with complete secure communications and surface search radar systems plus can be fitted with sonar and mine laying, detection and sweeping equipment if required. Normal armament is usually comprised of a 40mm Bofors gun on the forward deck, a 20mm Oerlikon cannon on the port side opposite of the pilot house, plus a variety of

AN OVERVIEW

weapons on the aft deck. These may include two .50-caliber M2 machine guns (on either aide), or a twin torpedo tube system, or an additional 20mm cannon aft, or a maritime direct fire 81mm mortar, which also mounts a .50-caliber machine gun.

	MK 3	**MK 4**
Overall length:	64 ft 10-3/4 in	68 ft 5 in
Overall beam:	18 ft 3/4 in	18 ft 3/4 in
Full displacement:	82,270 lbs	99,000 lbs
Draft:	5 ft 1 in	3 ft 1 in
Crew:	4	5
Passengers:	Limited	
Engines (shaft hp):	3 x 600 diesel	3 x 650 diesel
Propeller shafts:	2	2
Speed:	30 knots	30 knots
Range:		

MK 5 Patrol Boat (PUB). The Navy has released a classified request for proposals for a new PB, of which 15 are needed. Specifications require the capability to transport a SEAL platoon, transportable in a C-5A aircraft, be about 80 feet in length, and capable of 40 knots. One contender is the Israeli *Super Dvora* MK II patrol boat.

Cyclone Coastal Patrol Boat (PC). The Navy took delivery of the first of 13 of these steel-hulled boats in March 1993. They will replace the slower *Sea Spectres*, which are less capable in heavy sea states. The Cyclone can conduct operations for up to ten days and can keep pace with larger surface combatants. They are equipped with complete secure communications, surface search radar, and sonar systems. They mount two 25mm MK 38 chain guns (fore and aft), two .50-caliber and two M60 machine guns, plus are equipped with shoulder-fired Stinger air defense missiles. Future upgrades call for two stabilized Stinger mounts for up to six launchers. They will also receive some form of surface-to-surface missile system.

US NAVY SPECIAL WARFARE FORCES

Overall length:	170 ft
Overall beam:	
Full displacement:	
Draft:	
Crew:	28
Passengers:	10
Engines (shaft hp):	diesel
Propeller shafts:	2
Speed:	35 knots
Range:	2,000 nautical miles

River Raider Mini Armored Troop Carrier (MARC). The all-aluminum MATC was designed for high-speed patrol, interdiction, assault operations in rivers, harbors, and protected coastal areas. It is fitted with a short-range surface search radar and a multiple communications suite. The crew/troop compartment is lined with ceramic and Kevlar ballistic armor. The bow is fitted with a hydraulic ramp for troop insertion and extraction. Its flat bottom, shallow draft, and water jet propulsion make it ideal for river and inshore shallow water operation. The last MATCs were acquired in 1978. Seven weapons stations are fitted around the crew/troop compartment to mount any combination of M60 and M2 machine guns or MK 19 MOD 2 grenade launchers. Three MATCs can be airlifted in a C-5A transport.

Overall length:	36 ft
Overall beam:	12 ft 9 in
Full displacement:	27,390 lbs
Draft:	2 ft (1 ft at high speeds)
Crew:	2 (more required for weapons)
Passengers:	15
Cargo:	4,400 lbs
Engines (shaft hp):	2 x 280 diesel
Waterjet pumps:	2
Speed:	28.5 knots
Range:	370 nautical miles

MK 2 River Patrol Boat (RPB). The RPB was developed as a high-speed, highly maneuverable river craft for use in contested

AN OVERVIEW

areas. It can make a 180 degree turn in its own wake at full speed. Its fiberglass-reinforced plastic hull and water jet propulsion make it an excellent craft for use in shallow, debris-filled water. The coxswain is protected by ceramic armor. It is fitted with a short-range surface search radar and a multiple communications suite. Normal weapons complement include twin .50-caliber machine guns forward, two M60 machine guns (one on each side), plus an aft weapons station, which can mount a 60mm Mk 4 direct-fire mortar or 40mm grenade launcher or an additional .50-caliber or M60 machine gun. Four PRBs can be airlifted in a C-5A transport.

Overall length:	31 ft 11-1/2 in
Overall beam:	11 ft 7-1/2 in
Full displacement:	17,800 lbs
Draft:	2 ft
Crew:	4
Passengers	6
Cargo:	928 lbs
Engines (shaft hp):	2 x 210 or 280 or 300 diesel
Waterjet pumps:	2
Speed:	24 knots
Range:	200 nautical miles

Sea Fox Special Warfare Craft, Light (SWCL). The fiberglass hulled SWCL is a collapsible craft in the process of being phased out. It mounts radios, but no navigation aids. Armament includes various combinations of 7.62mm and .50-caliber machine guns plus a 40mm grenade launcher can be mounted.

Overall length:	35 ft 11-5/8 in
Overall beam:	9 ft 10 in
Full displacement:	26,000 lbs
Draft:	2 ft 10 in
Crew:	3
Passengers:	10
Cargo:	500 lbs
Engines:	1 diesel
Waterjet pumps:	1
Speed:	30 plus knots

US NAVY SPECIAL WARFARE FORCES

Range: 220 nautical miles

Small craft used by SEAL teams include the RIB, IRIB, and IBS plus various commercial inflatables, commonly referred to as combat rubber raiding craft (CRRC):

Rigid Inflatable Boat (RIB). The 24-foot RIB has a fiberglass reinforced hull and nylon/hypalon/neoprene sponsons. It features an inboard engine and center-mounted steering station. This small boat mounts radar, navigation, and communications systems. It can be transported on a trailer. A new RIB is in the process of being built by Novamarine for an eventual total of 72. This 10 meter craft will be able to deliver eight personnel at speeds up to 40 knots.

Overall length:	23 ft 9 in
Overall beam:	9 ft
Full displacement:	7,390 lbs
Draft:	A few inches
Crew:	1
Passengers:	15
Cargo:	2,800 lbs
Engine (shaft hp):	1 x 200 diesel
Propellers:	1
Speed:	
Range:	70 nautical miles

Interim Rigid Inflatable Boat (IRIB). The 30-foot IRIB has a fiberglass reinforced hull and trevira polyester sponsons coated with neoprene and hypalon. It features an inboard engine and aft-mounted steering station. This small boat mounts radar, navigation, and communications systems. It can be transported on a trailer.

Overall length:	29 ft 6 in
Overall beam:	10 ft 8 in
Full displacement:	14,700 lbs
Draft:	A few inches
Crew:	1
Passengers:	15

AN OVERVIEW

Cargo: 5,000 lbs
Engine (shaft hp): 1 x 300 diesel
Waterjets: 1
Speed:
Range: 150 nautical miles

Inflatable Boat, Small (IBS). The IBS has a neoprene hull and floor. It is fitted with a silent running outboard engine. The IBS is easily rigged for water parachute drop or water launching from the rear of a hovering CH-46 or CR-53 helicopter with full squad equipment secured on board.

Overall length: 12 ft
Overall beam: 6 ft
Crew: 1
Passengers: 8
Cargo: 1,000 lbs
Engine (shaft hp): 1 x 7.5 gas/oil

Commercial CRRC used by the SEALs include the Avon 450 (15 ft), 460 (15 ft), and 570 (17 ft); Z-Bird (15 ft), and Zodiac F-470 (15 ft). A choice of 15, 35, and 55 horse power outboard motors are available for these craft.

LIST OF ILLUSTRATIONS

FIGURE NUMBER		**PAGE**
Figure 1-1	The Mission Cycle	2
Figure 1-2	Phase Diagramming - Step One	6
Figure 1-3	Event Analysis - Step Two	9-11

CHAPTER 1

MISSION PLANNING PROCESS

1.0 INTRODUCTION

This chapter provides guidance for NSW mission planners. It includes discussion of a typical mission planning cycle; the use and benefit of the phase diagramming system (including an example); and a contingency planning checklist.

1.1 THE MISSION PLANNING CYCLE

A typical mission planning cycle is illustrated in Figure 1-1. It may be necessary to modify the order or delete one of the individual steps depending upon the available time, the operational commander's orders, or the nature of the particular mission.

1.2 THE MISSION PLANNING PROCESS

The mission planning process outlined in this section should be used as a general guideline and will require tailoring to fit individual missions.

1.2.1 RECEIVE THE MISSION DIRECTIVE.

1.2.2 INITIATE A SECURITY PLAN:
- Code name of operation.
- Security classification.
- Cover plans/stories.
- Identify personnel.
- Security measures.
- Formulate operational deception plan.

MISSION PLANNING PROCESS

- Isolation facilities.

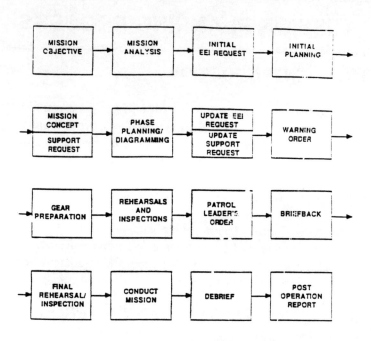

Figure 1-1. The Mission Cycle

1.2.3   ANALYZE THE MISSION:

- Review target analysis check list.
- Clarify exactly what the task is (specified and implied).
- Determine operational control (OPCON) and tactical control (TACON) over detachment during all phases of the mission.
- Determine if joint planning is required, and if so, availability of planning pubs such as Joint Operation Planning System (JOPS) Vol. IV. Additional pubs which will be of value are JCS Pubs 2, 6, and 20 and NWP 11.

- Ensure specific Rules of Engagement (ROE) are clear and not ambiguous.

1.2.4 PLAN THE USE OF AVAILABLE TIME (i.e., DRAW UP A TIME SCHEDULE).

1.2.5 SUBMIT AN INITIAL EEI REQUEST BASED ON MISSION ANALYSIS.

1.2.6 FORMULATE AN INITIAL PLAN:
- Assemble patrol members and review tasking(s).
- Study available intelligence.
- Make a thorough map/chart study.
- Review potential enemy weaknesses.
- Identify enemy strengths.
- Identify and assign relative values to the various elements of mission (i.e., surprise, speed, stealth).
- Consider limitations and special conditions regarding communications, logistics, support, communications security, movement, intel, and other requirements.
- Formulate several broad concepts of operations.
- Identify assets (i.e., what support will be required and its availability).

1.2.7 GIVE MISSION CONCEPT.
- Ensure concept contains or covers:
    ○ Variety - present more than one option.
    ○ Completeness - who, what, why, when, where?
    ○ Suitability - plan(s) which accomplish the assigned tasks.

MISSION PLANNING PROCESS

- ○ Feasibility - plan(s) can be accomplished with assigned or requested assets.
- ○ Rules of Engagement.
- ○ Acceptability - anticipated acceptable losses.
- ○ Limitations - operational limitations of plan compared to strength of your detachment.
- Request DIRLAUTH with supporting units.

1.2.8 REVISE PLAN, IF NECESSARY, BASED ON REVIEW OF MISSION CONCEPT.

1.2.9 PHASE PLAN/DIAGRAM THE MISSION TO IDENTIFY REQUIRED EEI, EEFI, REHEARSALS, TRAINING, EQUIPMENT, SUPPORT, AND POSSIBLE PROBLEM AREAS.

1.2.10 UPDATE EEI REQUEST IF NECESSARY.

1.2.11 SUBMIT SUPPORT REQUIREMENTS (AIRCRAFT, BOATS, FIRE SUPPORT, FREQUENCIES, CALL SIGNS, RESUPPLY, GASOLINE, ETC.).

1.2.12 ISSUE WARNING ORDER; BEGIN GEAR PREPARATION.

1.2.13 CONDUCT PRELIMINARY GEAR/PERSONNEL INSPECTIONS AND REHEARSALS.

1.2.14 UPDATE THE PLAN AS NECESSARY.

1.2.15 PATROL LEADER'S ORDER.

CHAPTER 1

1.2.16 BRIEFBACK.

1.2.17 FINAL INSPECTION, REHEARSALS, AND BRIEF.

1.2.18 CONDUCT MISSION.

1.2.19 DEBRIEF.

1.2.20 SUBMIT POST-OP REPORT.

1.3 THE PHASE DIAGRAMMING SYSTEM

Provided are the essential items which should be developed from the phase diagramming system during mission planning. Note that any mission can be planned using this system by substituting the events and critical problem areas for those used in the following example. It is essential that each event be critically analyzed for the three most likely causes of failure. Problem identification is necessary to conduct the training schedule prior to the actual operation to ensure discrepancies in material, personnel, or methodology do not cause real mission failure. These identified problems may even alter initial training techniques if they cannot be overcome during the early pre-operational training and rehearsals.

The phase diagramming of any mission may initially require six to eight hours of intensive concentration and brainstorming. This is common to all who have initially used this system. After a problem has been worked through, one should be able to finish planning a mission in two to four hours. The real key to the entire process and to a successful mission is the pre-operational training schedule and training objectives. Each event must be rehearsed (within given time constraints) to eliminate problems, check-out equipment, and mentally prepare personnel.

MISSION PLANNING PROCESS

1.4 ORGANIZATION OF THE PHASE DIAGRAM

Divide the mission into logical, independent phases and events. Although each mission will have its own unique profile, most missions can be separated into the following seven phases: pre-mission, insertion, infiltration, actions at the objective, exfiltration, extraction, and post mission. Figure 1-2 depicts an event breakdown of a rubber duck (CRRC paradrop)/long range transit/swimmer sneak attack mission with each phase broken down into various events.

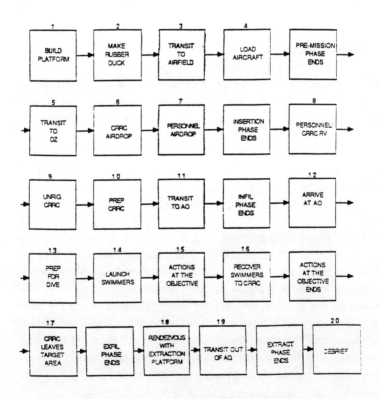

Figure 1-2. Phase Diagramming - Step One

CHAPTER 1

1.5 MISSION PHASES

The following are definitions of mission phases:
- Pre-mission. Ends when the insertion platform departs with the element onboard.
- Insertion. Ends when the element departs the insertion platform.
- Infiltration. Ends when the element reaches the objective area.
- Actions at Objective. Ends when the element departs the objective area.
- Exfiltration. Ends when the element boards the extraction platform.
- Post Mission. Begins at the time of return to the isolation area.

1.6 PHASE DIAGRAMMING

Phase diagramming is a method used to develop an operational plan. When complete, phase diagramming will:
- Confirm or deny the feasibility of a concept.
- Identify the most likely problem areas.
- Identify a complete list of EEI/EEFI.
- Identify all necessary equipment, rehearsals and support.

1.6.1 ANALYZE EACH EVENT AS FOLLOWS:

- Identify the three most likely things that can be wrong and three preventive or corrective actions (see Figure 1-3).
- Identify the earliest and latest likely times for beginning the phase, and the earliest and latest likely times for completing the phase (see Figure 1-3). By tracking the aggregate of earliest/latest beginning/completion times, it becomes immediately ap-

parent whether the mission can be completed in the available time.

1.6.2 CONTINUE THE ANALYSIS BY LISTING, FOR EACH EVENT:

- Necessary equipment
- Necessary EEI/EEFIs
- Necessary level of training/skills
- Necessary rehearsals
- Operational Security (OPSEC)
- Operational Deception (OPDEC).

1.6.3 PREPARE DETAILED LISTS.

After each phase has been thoroughly analyzed, complete lists of necessary equipment and EEI/EEFIs can be drawn up for preparation, and the lists of necessary skills and rehearsals can be checked against the list of things that can go wrong (Step 2). If time is a critical factor, training and rehearsals can be prioritized in accordance with their relationship to the list of things that are likely to go wrong (see figure 1-3).

CHAPTER 1

EVENT 11 TRANSIT TO AO Step Two
EEI Sea state/currents Weather/visibility Enemy surface/subsurface/air capabilities Time/bearing/distance to target Enemy patrols/shipping traffic Moon rise/set/phase Sun rise/set Enemy sensors/capabilities **EEFI** Fuel duration Friendly elements in area Transit time **SKILLS** OBM Mechanic Navigation Boat Maintenance/Boat Handling Experience in Long Boat Transits **TRAINING/REHEARSALS** Practice CRRC Long Range Navigation

Figure 1-3. Event Analysis - Step Two (Sheet 1 of 3)

MISSION PLANNING PROCESS

EQUIPMENT
(Based on a 14 man platoon; will require additional equipment for 16 man platoon)

ORD	Weapons as listed in Phase III Ammo as listed in Phase III
1st Lt	CRRC complete (extra gas lines, boat compass, tool kit, paddles, towing bridles, patch kits, etc.)
Intel	Maps/charts 14 compasses 2 binoculars
Comm/ ET	2 night vision devices 4 PRC-117 14 Motorola squad radios
Medical	Field med kit
Personnel	Life jacket/flare Mk 13/K-bar/cammied-up/ weapon/medical kit/E&E kit/compass/map (individual)/ web gear/ammo

TIMING

Earliest start time:	1900
Latest start time:	2100
Earliest completion time:	0300
Latest completion time:	0500

Figure 1-3. Event Analysis - Step Two (Sheet 2 of 3)

CHAPTER 1

THINGS THAT ARE LIKELY TO GO WRONG	PREVENTIVE/ CORRECTIVE ACTIONS
1. OBM breakdown	1. Trained mechanic. Bring tools and spare parts or spare motor, if space and time permits.
2. Navigational error	2. Study permanent geographical features and know tides, currents and winds. Take into account sea state. Always steer to left or right of target so once landfall is reached, a direction to target is established.
3. Run low on fuel or run out of fuel	3. Run trials with a fully loaded boat in various sea states to get exact fuel consumption figures. Take enough fuel for a worst case scenario.
4. OPSEC - Encounter enemy forces prior to mission completion	4. Execute immediate action drills upon contact; continue mission if possible or E&E.
5. OPDEC - Press corps penetrates Naval base after mission.	5. Avoid press, maintain cover story.

Figure 1-3. Event Analysis - Step Two (Sheet 3 of 3)

MISSION PLANNING PROCESS

1.7 PLANNING FOR CONTINGENCIES

The following is a list of situations that could negatively effect an operation. Some of these apply to every operation and should be planned for accordingly.

1.7.1 CONTINGENCY CHECKLIST.

One must formulate alternate plans to deal with the following potential contingencies:

- Launch or landing occurs at wrong position
- Date or time of the launch is delayed
- Late or early arrival at the objective
- Rough weather causes delay or cancellation of the operation
- Unpredicted tides/currents (tides or currents differing from those planned for the operational time period)
- Enemy contact (patrols, patrol boats, search lights, sentries, etc.)
- Discovery by local population (police, civilians, farmers, hunters)
- Craft and/or personnel separated or arrive at objective at different times
- Targets have diminished/increased or changed position
- More valuable targets are located
- Extreme weather or climatic change occurs
- One or more men become ill, injured, or die
- Craft capsizes, swamps, or is damaged beyond repair
- Element is captured by local civilians, police, or enemy forces

CHAPTER 1

- Member of element fails to reach one of your RVs (rendezvous sites)
- Lost anytime during operation
- Lose radio contact with base and/or another element
- Lose special equipment
- Planned resupply is a failure or cache is compromised
- Must dispose of equipment
- Escape and evasion plan must be initiated.

CHAPTER 2

ESSENTIAL ELEMENTS OF INFORMATION

2.0 INTRODUCTION

A thorough intelligence analysis directed at the requirements of the proposed mission will include an intensive review of the Special Operations Intelligence Folder (SOIF), if available, and will require that new intelligence and updates be provided by higher headquarters until the execution of the mission. If a SOIF is not available, the planners must immediately task intelligence support assets to provide an intelligence estimate based on the Essential Elements of Information (EEI) (Primary Intelligence Requirement [PIR] if from U.S. Army) list in Appendix A. Essential Elements of Information (EEI) are divided into two categories: target independent and target dependent. Target independent EEI are EEI which NAVSPECWAR forces would require regardless of the specific target in question. These EEI have been separated from the target dependent EEI to allow for more rapid processing of EEI submissions in the event of an actual mission assignment. The target independent EEI will be submitted immediately upon notification of mission tasking, even if the specific mission has not been identified. This will allow the intelligence community to begin producing intelligence in support of the mission as early into the mission planning process as possible. The target dependent EEI will be promulgated as soon as possible thereafter (i.e., once the specific target has been identified).

2.1 TARGET INDEPENDENT EEI

Target independent EEI's have been divided into the following four categories.

2.1.1 ENEMY ENVIRONMENT.

2.1.1.1 NATURAL OBSTACLES
- Topography.
- Meteorology.
- Hydrography.

2.1.1.2 MANMADE OBSTACLES

2.1.2 ENEMY ORDER OF BATTLE.
- Ground.
- Naval.
- Air.
- Communications.
- Electronic.
- Weapons.

2.1.3 SURVIVAL/EVASION/RESISTANCE/ESCAPE (SERE)

2.1.4 MISCELLANEOUS.

2.2 TARGET DEPENDENT EEI

Target dependent EEI's have been divided into the following categories:

2.2.1 IMAGERY AND GRAPHICS.

2.2.2 TEXTUAL DATA & SUPPORT MATERIALS.

CHAPTER 3

TARGET ANALYSIS

3.0 INTRODUCTION

Targets are not indiscriminately attacked. They are part of an overall plan to destroy an entire system. Interdiction is based on the assigned mission which directs, as a minimum, the results desired and the priorities of attack for specific systems. Based on this mission, the Patrol Leader selects the specific targets and those elements on which to conduct the attack. For specific Special Operations target vulnerabilities see references listed in Appendix F.1. Additionally, review the target analysis checklist in Appendix A.3.

3.1 TARGET SELECTION

Target selection requires detailed intelligence, thorough planning, and is based on the following six factors (CARVER).

3.1.1 CRITICALITY.

A target is critical when its destruction or damage will have a significant influence upon the enemy's ability to conduct operations. Each target is considered in relation to other elements of the particular target system designated for interdiction.

3.1.2 ACCESSIBILITY.

A target is accessible when it can be infiltrated either physically or by direct or indirect weapons fire (if and when that particular target can be destroyed/damaged by indirect fire methods).

3.1.3 RECUPERABILITY.

A target's recuperability is measured in time (i.e., how long will it take the enemy to replace, repair, or bypass the damage/destruction of the particular targets.

3.1.4 VULNERABILITY.

A target is vulnerable if the patrol has the means (i.e., explosives, weapons, manpower and expertise) to destroy or degrade the target.

3.1.5 EFFECT ON POPULACE.

Will the mission elicit a positive or negative reaction by the civilian populace? Will this reaction have long term effects if friendly forces move into the area?

3.1.6 RECOGNIZABILITY.

Will the target be easily recognized by the patrol? What features will assist in its identification?

3.2 TARGET SYSTEMS

A target system is a series of interrelated elements which together serve a common purpose. A target is one element; an installation, or an activity identified for attack such as a locomotive, a train, a bridge, or a prison. A target complex is numerous targets in the same general area such as a railway marshaling yard, an airfield, or port dock facilities. A target system may consist of an industrial system and its sources of raw material; the rail, highway, waterway, or airway systems over which these materials are transported; the source of power and method of transmission; the factory complex itself; and the means by which the finished produce is transported to the user.

TARGET ANALYSIS

3.3   MAJOR TARGET SYSTEMS INCLUDE:

3.3.1   RAILWAY SYSTEMS.

3.3.1.1 Railroad tracks are easily interdicted because it is almost impossible to effectively guard long stretches of track. Rolling stock may be simultaneously attacked with track interdiction. Loosening tie mountings, removing fishplates, offsetting track and using demolitions or special devices on curved sections of track or switches to cause train derailment, may result in captured or destroyed supplies, elimination of enemy personnel, or liberation of prisoners. Repair facilities and equipment are usually guarded and may be more difficult to attack.

3.3.1.2 Limited operations against railway systems and related facilities are only harassment, therefore, widespread operations are needed to severely effect the enemy.

3.3.2   HIGHWAY SYSTEMS.

Damaged highways are easily repaired and require less critical materials and skilled labor than railway systems. Therefore, points selected for interdiction should be in areas where the enemy cannot easily re-establish movement by making a short detour. Since highways have fewer vulnerable spots, these critical points will likely be heavily defended. Where highways cannot be destroyed traffic can be disrupted by successive roadblocks, real and dummy mines, booby traps, sniping, misdirection of route signs, or by spreading objects for puncturing tires. Ambushes are conducted when suitable terrain is available.

3.3.3   WATERWAY SYSTEMS.

The most critical facilities of waterway systems are ports, dams, canals, locks, and related repair equipment. They are usually well guarded since their destruction can disrupt water traffic for long periods. Waterway control and navigational equipment such as signal lights, beacons, microwave communications systems, channel markers and buoys can be attacked effectively. Sinking vessels in restricted channels, dropping bridges in waterways, creating slides, and destroying levees can block waterway traffic. Because of security and the amount of explosives required, destroying a dam will often be beyond the capability of small units. A lesser degree of damage (i.e., destroy floodgates, sluice gates, etc.), is an option.

3.3.4 AIRWAY SYSTEMS.

The enemy's military/commercial airway systems can be disrupted by interdicting airfields, parked aircraft, and related facilities. Terminals, hangars, repair shops, field depots, radar and radio navigation controls, lighting, communications, and defense systems are targets. Eliminating flight and ground personnel is also an option. Weapons may be available to attack and destroy low flying aircraft.

3.3.5 COMMUNICATION SYSTEMS.

Widely dispersed communication systems present excellent targets. Cutting telephone wires, damaging telephone terminals, destroying microwave antennas, or destroying transmitters usually results in degradation of communications. Alternate and emergency means of communication are usually available, however, destruction of any part of a communication system creates an overload on remaining facilities.

3.3.6 POWER SYSTEMS.

Electrical power nets can be interdicted by destroying cross-country or local high tension lines. Distribution power lines located in remote areas, which would make repair/replacement difficult are good targets. Substations, although critical, can be bypassed in a relatively short time by improvised wiring. Interdiction of power nets can be accomplished by destroying power generating stations and related equipment.

3.3.7 WATER SUPPLY SYSTEMS.

Water systems supplying industry can be disrupted by attacks against reservoirs, pipelines, and purification plants.

3.3.8 FUEL SUPPLY SYSTEMS.

Attacks against an enemy's fuel supply system have far-reaching effects on his economy as well as his ability to conduct and support military operations. Surface and underground fuel storage tanks, depots, pipelines, refueling systems for tank trucks, rail tank cars, transport vehicles, and vessels are all good targets.

CHAPTER 4

MISSION CONCEPT

4.0 INTRODUCTION

Once mission and target analysis have been completed, the Patrol Leader should formulate his basic concept of operations prior to starting his detailed planning. The Patrol Leader presents the Operational Commander with his concept. This permits the Operational Commander to make recommendations, alterations, and/or refinements to the concept prior to his final mission planning. As this is only a presentation of the Patrol Leader's concept of mission execution, general rather than specific details will be presented in the brief. Upon completion of the mission concept, the Patrol Leader should query the Operational Commander concerning any specific questions not yet defined under Rules of Engagement.

4.1 MISSION CONCEPT (FORMAT)

Upon completion of initial planning, a message is sent in the following format to the Operational Commander (if not co-located) to allow for coordination of the mission.

A. **Mission.** Mission statement taken directly from the mission directive

B. **Insertion.** Method, time, equipment, and support required

C. **Infiltration.**
- Method, time, and equipment
- Route to target, i.e., beach landing site, DZ or water DZ, LUP's (Lay Up Points) en route

MISSION CONCEPT

D. **Execution.** (actions at objective)
- Type of action at the objective
- Method of reconnaissance of targets
- Ordnance/demolitions to be used
- Security during action at the objective
- Action upon completion of target execution
- Preplanned fire support plan

E. **Exfiltration.**
- Method, time, equipment and support required
- Route from target
- LUP's en route

F. **Extraction.** Method, Time, equipment and support required

G. **Alternative Actions.** Plan for rendezvous of element and subsequent action in case of compromise or escape and evasion.

H. **State Assumptions.** (i.e., infil/exfil platforms available)

I. **State Support Requirements.** (i.e., aircraft, fire support, etc.)

J. **State Operational Limitations.**

4.2 RULES OF ENGAGEMENT (ROE) CONSIDERATIONS

The Patrol Leader must know precisely what the priority of the mission is before departing isolation, in terms of:

A. Is the mission continued or aborted if:
- **Suspected** compromise occurs by civilians or military.

CHAPTER 4

- **Known** compromise occurs by civilians or military.
- Contact has been made and broken
- POWs/detainees have been taken
- An indigenous guide refuses to continue at some point in the patrol.

B. Degree of aggravation among the local populace is acceptable.

C. Primary target has been altered or degraded.

D. What is highest priority: remaining covert, clandestine, or taking out target regardless of cost?

E. Alternate targets/mission. Are there secondary missions in case primary cannot be fulfilled? Are there higher priority targets/missions of opportunity which may take priority over assigned mission?

CHAPTER 5

WARNING/PATROL LEADER'S ORDERS

5.0 INTRODUCTION

This chapter discusses the use of both the Warning Order and the Patrol Leader's Order (PLO). It identifies the preparatory intent of both orders while demonstrating the greater level of detail required to issue an effective PLO. Format samples for both orders are included in each respective section.

5.1 THE WARNING ORDER

The Warning Order is to warn the patrol members of an impending mission and to organize their preparation for that mission. The format outlined below covers the information necessary for a warning order. The detail covered in each section is determined by the Patrol Leader to ensure proper understanding by his patrol.

5.2 WARNING ORDER (FORMAT)

A. **Situation:** Brief statement of enemy and friendly situations

B. **Mission:** State in a clear and concise manner the mission of the patrol (use the tasking message as your basis)

C. **General Instructions:**

1. State the general and special organization to include the element or the team organization and the individual duties. For each person state:

CHAPTER 5

- Position or primary responsibility
- Weapons, ammunition and demolition material (type and quantity) to be carried
- Assignments to special detachments or teams
- Special individual equipment
- Assignments for preparing platoon equipment (type, quantity, and expected operational requirements).

2. The uniform and equipment common to all, to include:
    - Type of uniform
    - Civilian, deceptional clothing, or camouflage
    - Web gear
    - Escape and evasion gear
    - Footwear
    - Rations
    - Water
    - Sleeping gear.

3. Weapons, ammunition, and equipment each member will carry.

4. Chain of command.

5. A time schedule for the patrol's guidance on:
    - Drawing equipment
    - Test firing of weapons
    - Muster(s)
    - Patrol Leader's order
    - Support personnel brief
    - Departure.

WARNING/PATROL LEADER'S ORDERS

6. Time, place, uniform, and equipment for receiving the patrol order.

7. Times and places for inspections and rehearsals.

D. **Specific Instructions:**

1. To subordinate leaders

2. To special purpose teams or key individuals.

5.3 RECOMMENDED BRIEFING ASSIGNMENTS

Although one should be capable of writing the warning order alone, it is better to involve the platoon in the process. One needs to review inputs, and their research will save considerable time. Recommend the following assignments:

a. Situation/graphs/charts	Intelligence Representative
b. Dive brief	Diving Supervisor
c. Jump brief	Jumpmaster
d. Cast brief	Castmaster
e. Navigation	Navigator (Pointman)
f. Communications plan	Communicator
g. Medical plan	Corpsman
h. Escape and evasion plan	Intelligence Representative

5.4 PATROL LEADER'S ORDER

The Patrol Leader's Order (PLO) is used to pass the detailed plan to those tasked with execution and selected others who need to know. Phase diagramming is the preferred method used to develop operational plans leading to issuance of a PLO. The success of an operation may be directly attributed to the quality of the orders the Patrol Leader issues for the operation.

5.5 PLO SEQUENCE

A set PLO sequence is used to ensure that:
- All relevant information is included.
- It is logical.
- It is clear and concise.
- It is easy to follow for taking notes.
- It is easy for recipients to quickly grasp all details.

5.5.1 THE FOLLOWING FIVE PARAGRAPH SEQUENCE IS USED:

- **Situation.** Hydrographies, topographies, weather, and intelligence data. What is going on now and developments which have led up to the present situation.
- **Mission.** What is the task?
- **Execution.** How will task be performed?
- **Administration and Logistics.** Administrative requirements for the task.
- **Command and Signals.** Command and communications aspects.

5.5.2 BAD PLOs.

If the recipients of your orders know why the plan was made, what the intent is, how, when, and where to carry the orders out; and what part the individuals are going to play; your orders will be successful. Bad PLOs are characterized by:
- Confusion
- Lack of confidence in you or your plan
- Failure to carry out the task.

WARNING/PATROL LEADER'S ORDERS

5.6 PRE-PLO CHECKLIST

Prior to the arrival of the PLO recipients ensure that:
- The location of the brief is prepared (seating, boards, lighting, etc.).
- A model or sand table of the objective is made up, when possible.
- Maps, charts, air photos, and other aids are available. They should remain out of sight until they are presented.
- Set security in the briefing area.
- Conduct a muster.
- Seat all the personnel in a logical sequence (e.g. swim pairs, assault groups, support personnel, etc.) - no one should be allowed to lie down, sit in the back of the room, etc.
- Ensure that all personnel have the necessary materials required to receive the orders (e.g., pencil and paper).

5.7 PRESENTATION

Ability to give a good Patrol Leader's Order depends upon subject knowledge and presentation techniques. Consider yourself a salesman while giving your PLO. When finished, everyone should be "sold on the idea." Remember, PLOs are orders and not a planning conference nor a time to debate tactical options. If you do not know the answer to a question, admit it. **DO NOT BLUFF.**

5.7.1 BRIEFING TECHNIQUES.

Present the PLO clearly and concisely, ensuring that:
- All headings from the PLO card are given.

- State that questions will be taken after each major section (i.e., situation, mission, execution, administration and logistics, command and signals).
- State the mission twice.
- State all coordinates twice (other than in the mission which is given twice anyway).
- Use a pointer for charts, models, and photographs. When necessary, have someone assist by pointing as you brief.
- Synchronize watches at the end of the PLO, then very generally summarize plan and allow time for all personnel to consider the orders.
- Take questions from the members of the group to help clear up points of confusion.
- **Ask** confirming questions.

5.7.2 PLO INTRODUCTION.

A. Muster

B. Set security

C. Time check

D. Warning order review (ensure all assigned tasks have been completed)

E. Mission (brief statement including reason for the tasking)

F. Chain of command and description of duties.

WARNING/PATROL LEADER'S ORDERS

5.8 PLO FORMAT

5.8.1 SITUATION.

A. **Weather**
- Visibility
- Wind
- Weather
- Temperature
- Precipitation
- Cloud cover
- Water temperature
- Sun rise and set
- Moon rise and set
- Tides:
    - Low_____
    - High_____
- Current
- Surf:
    - Height_____
    - Period_____

B. **Terrain**
- Type of terrain
- Relief
- Vegetation
- Density of vegetation
- Cover
- Concealment
- Roads
- Rivers, canals, streams on routes

- Clearing for LZ's
- Population concentrations
- Enemy installations on routes
- Obstacles (swamps, bogs, cliffs, etc.)
- Suitability for radio transmission
- Overhead canopy
- Beach
    - obstacles
    - gradient
    - current
    - kelp
    - width/depth of beach
    - type of sand
    - trafficability
    - beach exits
    - hinterland vegetation
- Drinking water availability
- Aerial photos/maps available

C. **Enemy**
- Identification
- Location
- Activity
- Strength
- Clothing
- Weapons
- Emplacement/fortifications
- Warning systems
- Domestic animals
- Booby traps or mines

WARNING/PATROL LEADER'S ORDERS

- Estimate of action on contact
- Routes, modes, and times of travel
- Enemy force activity/routine.

D. Friendly
- Transportation available
- Fire support available
    - how much
    - what kind
    - reaction time
    - accuracy
    - spotting method
    - reliability
- Resupply sources available
- Other friendly patrols
    - how many
    - where
    - identification
    - mission
- Mission of next higher unit
- Guide availability

E. Target
- Location
- Obstacles
- Natural defenses
- Illumination
- Avenues of approach
- Best method of finding target
- Number and types of structures

F. **Other**
- Civilian attitude toward U.S. military
- Economic situation of population
- Education/cultural factors
- Religion

G. **Reliability of Intel Source**

5.8.2 MISSION:

What the patrol is going to accomplish and the location or area in which it is going to be done.

5.8.3 EXECUTION.

A. **Overall concept**

B. **Other Missions**

C. **Coordinating Instructions**
- Time schedule (WO)
- Primary insertion
    - time schedule
    - location
    - method
- Positions in insertion platform
- Primary approach route
- Departure for friendly areas
    - identification
    - location
    - method

WARNING/PATROL LEADER'S ORDERS

- Prominent terrain/manmade features along approach route
- Organization for movement during approach
- Actions at danger areas
    - rivers
    - roads and paths
    - open areas
    - built-up areas
- 9. Rallying points
    - IRP (Initial Rally Point)
    - ORP (Operational Rally Point)
    - others
- Actions at the objective area
- Organization for movement during exit
- Primary exit route
- Prominent terrain/manmade features along exit route
- Re-entry into friendly areas
    - identification
    - location
    - password
    - signals
- Primary extraction
    - time window
    - location
    - method
- Positions in extraction platform
- Debriefing

CHAPTER 5

D. **Alternate Plans and Contingencies**
- Alternate insertion/extraction
    - insert
        - (1) time window
        - (2) location
        - (3) method
    - extract
        - (1) time window
        - (2) location
        - (3) method
- Alternate routes
    - approach
    - exit
- Drop dead/turn around times
- Actions on enemy contact
    - ambush (sound off, return fire)
        - (1) front
        - (2) flank
        - (3) rear
        - (4) on insertion/extraction
        - (S) when patrol is split
        - (6) crossing stream or road
        - (7) in boat/helo/CRRC/vehicle
    - casual contact
    - ineffective/random fire
    - booby traps
- Handling wounded/dead
- Escape and evasion plan
- Other

WARNING/PATROL LEADER'S ORDERS

5.8.4 ADMIN AND LOGISTICS.

A. **Rations/water**

B. **Arms/ammo** (WO)

C. **Uniform and Equipment** (WO)

D. **Special Equipment** (WO)

E. **Resupply Plan:**
- Time _____
- Source
- Supplies
- Signals

F. **Handling Wounded**

G. **Handling Prisoners**
- Search, separate, silence, speed, safeguard
- Retain all items found
- Field interrogate
- Life jacket
- Handling instructions

5.8.5 COMMAND AND SIGNALS.

A. **Hand Signals**
- Stop
- Set perimeter
- Danger area
- Head count
- Pace
- Enemy

CHAPTER 5

- Friendly
- Hear something
- See something
- OK
- Get down
- Speed up
- Slow down
- Open interval
- Manmade structure
- Booby trap
- 17. Rally
- Road
- Objective
- Get on line

B. **Radio Communications**
- Frequencies
    - primary
    - secondary
    - admin
    - emergency
- Call signs
- Codes/code words
    - insert
    - extraction
    - shift frequencies
    - contact
    - medevac
    - shore bombardment/artillery support
    - air support

PATROL LEADER'S HANDBOOK

WARNING/PATROL LEADER'S ORDERS

- ○ cease fire
- Authentication plan
- Time/type of reports

C. **Challenge and Passwords**

D. **Lost Comm Plan**

E. **Position Marking (Day and Night)**

F. **Enemy Position Marking**

G. **Command**
- Chain of command
- Location of Leaders
    - ○ during insertion/infiltration
    - ○ in patrol
    - ○ in danger areas
    - ○ at objective
    - ○ during extraction/exfiltration

CHAPTER 6

BRIEFBACK

6.0 INTRODUCTION

A briefback is a detailed brief given by the Patrol Leader, and if required, key members of the patrol, to the Operational Commander for the purpose of demonstrating to him that:

- The operational plan is well thought out and complete.
- The members of the patrol are familiar with the plan and understand their role in the operation.
- The plan will accomplish the assigned objectives.

A briefback is given near the end of the planning cycle, after the entire plan has been developed. The exact location of the briefback in the planning cycle, as well as the format and amount of detail required, will depend upon the Operational Commander. The following format is an **example** of a detailed briefback. The brief has been divided into sections and may be given by various members of the patrol. This is a method that **may** be used to accomplish the objectives stated above.

Briefbacks may or may not be required depending on who the Operational Commander is, where he is located relative to the platoon, and how much time is available prior to mission execution. Additionally, the situation may dictate a briefback lasting from 10 minutes to over an hour. The platoon commander is provided the following information in order to familiarize himself with the briefback.

BRIEFBACK

6.1 BRIEFBACK (FORMAT)

6.1.1 SITUATION/MISSION - Briefed by Patrol Leader.

A. **Classification.**

B. **Overall situation** - any changes or updates from target folder.

C. **Mission** - as stated by tasking (state twice).

D. **Purpose.**

E. **Assumptions and operational limitations.**

6.1.2 INTELLIGENCE - Brief *threat* to detachment.

A. **Area of operations.**
- Weather
    - Existing situation. Include light data and climatic information or a weather forecast, significant to the mission.
    - Effects of the weather on the friendly situation to include effects on reaction time and courses of action.
    - Effects of the weather on the friendly situation to include effects on personnel, equipment and actions.
- Terrain.
Relate the following factors to the mission and explain the effects on both the enemy and friendly situation.
    - Observation and fields of fire.
    - Cover and concealment.

- Obstacles (manmade and natural).
- Key terrain features.
- Avenues of approach available to both enemy and friendly forces.
    (1) High speed routes to infiltration, target and exfiltration areas.
    (2) Effects on enemy reaction time and ETA for each route.
* Other characteristics, include only the information which may have an effect on the mission (i.e., transportation systems, hydrography, communications systems, etc.).

B. **Enemy situation.** The general description of the enemy situation to include details of enemy forces which may effect the mission. Be concise.

* Disposition (reference an overlay).
* Composition and strength.
* Committed forces and reinforcements.
  Compute enemy reaction time to the objective areas, LZ/DZ/BLS, and the rally points. Relate this information to the time schedule.
* Other enemy capabilities (tactical air support, air movement aircraft, CBR capabilities, RDF equipment, etc.).

C. **Friendly situation**

D. **Other intelligence factors as they relate to the mission.**

* Recent and present significant activities of the civilian populace (curfews, population control measures, etc.).
* Peculiarities and weaknesses which may affect the mission: personnel, intelligence, operations, com-

BRIEFBACK

> bat service equipment, civil-military operations and personalities.
> - EEI.
> - Intelligence reports, as required.
> - Map coverage.
> - Counterintelligence measures.
> - Estimates of guerrilla forces and underground organizations. Include the following: disposition, composition, capabilities, recent and present significant activities, peculiarities and weaknesses strength, leadership, morale, and security measures.

6.1.3 GENERAL OVERVIEW.

A. **Concept of operations.**

B. **Unit organization and chain of command.**

C. **Personnel Responsibilities.**

6.1.4 INSERTION METHOD.

6.1.5 ROUTES.
- Infiltration.
    - Primary.
    - Alternate.
- Exfiltration.
    - Primary.
    - Alternate.

6.1.6 ACTIONS AT THE OBJECTIVE - BRIEFED BY THE PATROL LEADER.

A. **Target orientation.**

B. **Target analysis.**

C. **Method of attacking target.**

D. **Alternate plans.**

6.1.7 EXTRACTION METHOD.

6.1.8 RENDEZVOUS/EVASION AND ESCAPE PROCEDURES.

A. **Rally points and rendezvous plans.**

B. **E&E plan for all phases of the operation.**

6.1.9 COMMUNICATIONS.

A. **Equipment.**
- Type.
- Quantity.
- Comm plan.

C. **Lost comm plan.**

D. **Internal communications.**

E. **CEOI considerations.**

BRIEFBACK

6.1.10 MEDICAL.

A. **Health status of detachment.**
- Shots up to date.
- Medicinal requirements.

B. **Medical training of detachment.**

C. **Precautions/preventative measures.**

D. **Handling of injured.**

E. **Nearest friendly medical facility.**

6.1.11 CLOSING STATEMENTS - BRIEFED BY PATROL LEADER.

A. **Readiness of detachment.**

B. **Questions.**

C. **Classification.**

6.2 BRIEFBACK PRESENTATION

A briefback presentation should be prepared to last about one hour, with about forty minutes devoted to briefing various phases and about twenty minutes for question/answer and discussion. Briefback presentations should include the following:

A. **Briefback Packet.** Present Operational Commander and key personnel present with a packet of your final plan, including phase diagrams. The packet should be neatly written or typed double-spaced in large script and easy to read. The brief should follow along with the packet, allowing each person to read the brief's high points, see the applicable maps/routes, and easily write down notes or questions. A copy of This packet will remain with the Operational Commander and Mission Coordinator, Should questions arise concerning the extraction plan, RV plan or other crucial phases of the mission while the platoon is in the field.

B. **Visual Aids.** Key visual aids should be included in the briefback packet. The only remaining necessary visual aids are maps and pictures of the AO from small to large scale and a general time schedule. Such visual aids should be referred to a number of times during the course of the brief.

C. **Purpose of Mission.** Briefly explain the significance of the mission to future operations (e.g., how the radar installation you are disabling will allow a follow-on air strike).

D. **Situation Brief.** This brief should highlight the natural disadvantages and enemy strengths the patrol will avoid and, conversely, the natural advantages and enemy weaknesses the patrol will exploit.

E. **Assumptions.** State the areas where key intelligence was available and the logical assumptions made based on the intelligence that was available.

F. **Execution Decisions.** With every phase of the operation, briefly explain what the Patrol Leader is doing and key decisions expected during the operation.

G. **Operation Weaknesses.** Every operation has a weak phase or phases. Identify these phases as such, while clearly outlining precautions taken and back-up plans to minimize chances of failure.

H. **Detailed Plans.** As a rule of thumb, briefbacks should brief the platoon's plans in general and coordinated support requirements in detail (e.g., brief where and when helicopter pickup required, not how the patrol will sit in the helo). Detailed platoon plans should be covered in the PLO during isolation.

BRIEFBACK

I. **Insertion/Extraction.** Alternate RV plans. Those should be planned and coordinated in detail and simply briefed (time, place, signals and method).

J. **Infiltration/Exfiltration Modes.** These should be planned in detail and simply briefed. For example, the exact course(s) and seating arrangement need not be briefed for a CRRC infiltration, but highlights such as overall distance, best worst speeds and anticipated fuel consumption should be briefed.

K. **Time Line/Phase Diagram.** Phase diagram should be included in the briefing packet and should be used during brief and execution phase. Emphasize key phases, alternatives, drop dead times and related information.

L. **Communications.** Include a complete CEOI but brief only the highlights. Also include a "no comms" plan of action.

M. **Question and Answer Session.** As discussed, brief the highlights but have all the details of PLO available should questions arise.

N. **At the Conclusion of the Briefback:**
- Security.
  Upon conclusion of brief, collect all notes/briefing material including all briefback packets and leave them with Mission Coordinator or Operational Commander.
- Sensitive Equipment.
  Serial numbers or copies of Forms DD1149 for weapons, radios and other sensitive equipment should be left with the Mission Coordinator or Operational Commander prior to departure on the mission.

CHAPTER 7

POST EXERCISE/OPERATION REPORTS AND INTELLIGENCE DEBRIEFS

7.0 INTRODUCTION

All Post-Exercise/Operation reports are to be prepared in accordance with the following format.

7.1 POST-EXERCISE/OPERATION (FORMAT)

From: Officer-in-Charge, _____ Platoon, SEAL Team ___

To: (Operational Commander)

Via: (Chain of Command)

Subj: POST-OPERATION REPORT FOR:

Ref::

Encl: To include but not limited to:
- Activity participants
- Schedule of key events
- Details on any subject which the writer wishes to treat in depth (e.g., proposed new procedures, details of a certain aspect of an operation, analysis of OPAREA, etc.).

POST EXERCISE/OPERATION REPORTS

- Comments and recommendations section from the past after action report for this activity. This will provide the reader with a perspective of the evolution being conducted, whether recurring problems limit its value, and whether there is an upward or downward trend in the value of the training.
- TIMS/EPS Abstract
  - Background.
    Key references and events which had a significant impact on the activity and its outcome.
  - Summary.
    A summary of what happened. This should be brief. Items that the writer believes warrant detailed treatment should be dealt with in an enclosure.
  - Comments and Recommendations by Topic:
    (1) Comment. As appropriate.
    (2) Recommendation. As appropriate.
    (3) Action. If the comment warrants a recommendation, this subparagraph should identify who the writer believes should take action on the recommendation.

7.2 INTELLIGENCE DEBRIEF GUIDE

The enclosed general intelligence debriefing guide should be reviewed by the Patrol Leader prior to an operation in order to enable him and other members of the patrol to collect needed intelligence. An intelligence debriefing will normally be conducted within a few hours of the completion of the operation. Because this is a general intelligence debriefing, some of the topics discussed may or may not be applicable to your operation.

7.3 DEBRIEF GUIDE (FORMAT)

A. Route and outline time frame.

B. Task required and details of how and if accomplished.

C. Contact:
- Where and when and who fired first.
- Enemy strength.
- Description - race - sex - dress - equipment - weapons - any known faces - ranks.
- Action - what they were doing - direction of movement - reaction to contact.
- Casualties - own men - what happened/what was done with enemy bodies.
- Evidence recovered - documents - equipment - weapons, etc.

D. Sighting. Same as for a contact but in addition:
- How many patrol members sighted.
- What details were seen.
- What evidence was left of what you were doing.

E. Spottings. Relevant headings of contacts.

F. Tracks.
- Location and direction.
- Age.
- Number of personnel using trail.
- Estimated destination/origin.

G. Aircraft/ships/vehicles.
- Where and when.
- Direction of heading.

POST EXERCISE/OPERATION REPORTS

- Altitude/speed.
- Number.
- Identification.
- Miscellaneous.

H. Camps found.

- Location and description of terrain.
- Size.
- Enemy strength in it or using it.
- Radio sets/aerials.
- Enemy activity.
- Structures - type, number, age.
- Fortifications, booby traps, dug outs, etc.
- Obstacles.
- Sentry arrangements and warning signals.
- Possible escape routes and approaches - direction of.
- Food dumps in camps.
- Weapon and ammo dumps.
- Printing presses.
- Documents.
- What was done to the camp.
- Miscellaneous.

I. Supply dumps found.

- Location and time of discovery.
- Contents.
- Condition.
- How concealed.
- When last visited.
- Age.

- Added to since first laid down.
- What was done to the dump.

J. Cultivation areas.
- Location.
- Size and shape.
- Any steps taken to camouflage it.
- Type of crops.
- Age.
- When last tended.
- Any signs of habitation in area.
- Any tracks.
- If near native settlement - estimate excess over local demand.

K. Local people (from known location)
- Location.
- Village of origin.
- Name of tribe and headman.
- Number.
- Friendly.
- Contact previously with armed forces.
- Moved recently. If so, why.
- Any information given.

L. Topography.
- Intel brief accurate. If not, what were inaccuracies.
- Map accurate. If not, what were inaccuracies.
- If air photos used, was the interpretation correct and useful.
- State of tracks, if used.

POST EXERCISE/OPERATION REPORTS

- Had tracks been recently used.
- Any other tracks or game trails found. If so, where.
- Rivers:
  - depth, width and speed
  - bridges
  - fjords.
- Sea:
  - sea state
  - currents
  - tidal range, etc.
- Water points.
- Laying up points (LUP).

M. Equipment.

N. Rations.

O. Morale/welfare.

P. Health.

Q. Security. (i.e., if not sighted by enemy or locals, any traces of lost equipment left behind which might be found later - if sighted, was your position or action likely to indicate future or other friendly activity).

R. Administration:
- Did you have adequate preparation time and facilities.
- Any comments on the support from base during the operation.
- Any equipment lost.
- 4. Anyone not likely to be fit for immediate further employment.

APPENDIX A

NSW INTELLIGENCE

A.1 TARGET INDEPENDENT EEI - ENVIRONMENT

A.1.1 OBSTRUCTIONS/CONSTRUCTIONS

A. **Natural Obstructions**
- Topographic EEI
- Meteorological EEI
- Hydrographic EEI.

B. **Manmade Constructions/Obstructions**

A.1.2 ORDER OF BATTLE (OOB)

A. Ground OOB

B. Naval OOB

C. Air OOB

D. Communications OOB

E. Electronics OOB

F. Weapons OOB.

A.1.3 SERE

A. Evasion/Escape Routes

NSW INTELLIGENCE

    B. Pertinent Cultural Considerations

    C. Enemy Counter-Evasion/Escape Tactics

    D. Contact Plans.

A.1.4 MISCELLANEOUS INFORMATION

A.2 TARGET DEPENDENT EEI

A.2.1 IMAGERY AND GRAPHICS

    A. Area Orientation Imagery

    B. Target Orientation Imagery

    C. Target Imagery

    D. Target Graphics.

A.2.2 TEXTUAL DATA/SUPPORT MATERIALS

    A. Target Description

    B. Target Area Activity

    C. Enemy Reaction Capability.

A.3 TARGET ANALYSIS CHECKLIST

A.3.1 ADMINISTRATIVE DATA

    A. Name of Facility

    B. Location (Address)

    C. Date of Analysis

APPENDIX A

D. Author and Sources

E. List of Attachments.

A.3.2 GENERAL

A. Facility Description

B. Facility Component Parts.

A.3.3 SPECIFIC

A. Potential Target List

B. Common Target List

C. Target Relationship to Support Facilities.

A.3.4 CONCLUSIONS

A. Target Attack Profile

B. Target Damage Estimate

APPENDIX A.1

NSW INTELLIGENCE

A.1 TARGET INDEPENDENT EEI

A.1.1 OBSTRUCTIONS/CONSTRUCTIONS

A. **Natural Obstructions**
- **Topographic EEI.** Topographic characteristics in the area of operations that would be favorable or limit the successful execution of a Naval Special Warfare mission.
    - Natural obstacles (i.e., mountains, cliffs, swamps, etc.)
    - Paths/trails (i.e., type, location, directions, purpose, dimensions, etc.)
    - Estuaries (i.e., waterways, rivers, streams) to include type, direction, depth, location, presence of rapids, drainage systems, etc.)
    - Hazardous areas/open plains/snow fields, etc.
- 2. **Meteorological EEI.** Meteorological characteristics in the area of operations that would be favorable to or limit the successful execution of a Naval Special Warfare mission.
    - Atmospheric forecasts including:
        (1) Wind direction and speed at all altitudes up to 30,000 feet
        (2) Sky conditions (e.g., dry/wet, percentage cloud cover, presence/locations/speed/direction of storm centers, etc.)

APPENDIX A.1

  (3) Air temperature up to 30,000 feet
  (4) Weather extremes for the area
  (5) Humidity percentage
  (6) Effect of the topology on the weather
  (7) Presence/effect of sunspots
  (8) Presence/effect of electrical interference
  (9) Presence/effect of fog or mist.
- Tabular data for
  (1) Sunrise
  (2) Sunset
  (3) Beginning of Morning Nautical Time (BMNT)
  (4) End of Evening Nautical Time (EENT)
  (5) Moonrise
  (6) Moon phase
  (7) Percentage of illumination.
  (c) Star data for the area
- Procedures for acquiring most accurate climatological forecasts for:
  (1) Next 24 hours
  (2) 24-36 hours
  (3) 36-72 hours
  (4) 72+ hours.

- **Hydrographic EEI.** Hydrographic characteristics in the area of operations that favor or limit the successful execution of a Naval Special Warfare mission.
- Water temperature, to include thermocline layers
- Speed, direction and schedule for currents
- Direction, range, and schedule for tides

NSW INTELLIGENCE

- Bioluminescence data for the area
- Depths for all water in the area
- Water surface conditions, including floating or stationary ice
- Debris on the water surface in the area
- Coastal gradients in the area
- Location and nature of any breakwaters in the area
- Bottom composition
- Water turbidity factor
- Salt water intrusion from the sea/ocean into inland water
- Dangerous marine life
- Location and nature of any submerged natural obstacles (e.g., coral reefs)
- Location and nature of any submerged manmade obstacles or objects (e.g., wrecks, pipelines, cables, etc.).

B. **Manmade Constructions.** Manmade facilities/constructions in the area which impact on the success of a Naval Special Warfare mission.

- Locations, dimensions, construction and functions of all civilian and military facilities in the area (e.g., buildings, water towers, power stations, roadways, airfields, rail lines, bridges (highway and railroad), tunnels, external visible lighting)
- Civilian and military populations housed in the area
- Locations, operation, and function of any subsurface water intakes (e.g., for water purification or hydroelectric plants) in the area
- Land and water navigational aids in the area and their functions
- Fences/barricades/mine fields/sensor fields.

A.1.2 ORDER OF BATTLE

A. **Ground OOB**
- Organized national army ground forces located in the area.
    - Designation
    - Location
    - Manning level
    - Morale
    - Level of training
    - Uniforms
    - Current activities
    - Combat effectiveness
    - Missions and functions
    - Capabilities
    - Operational limitations
    - Equipment (tanks, vehicles, weapons, etc.).
- Organized national paramilitary forces located in the area.
- Special national forces located in the area (e.g., militia, police, youth groups, terrorists, KGB/GRU-type forces, local defense forces, coast watchers).
- National guard force reaction capability located near the area.
    - Reaction time
    - Avenue of approach into the area.
- Forces from other nations, especially from the Soviet Union, in the local area.
- Ground force active/passive defensive measures employed in the area.
    - Guardposts, watch towers, checkpoints, or security stations, where located and how

NSW INTELLIGENCE

manned. Defensive precautions, stages of alert, procedures when alerted. Guard rotation schedules.
  - Ground patrol routes, patterns, schedules, etc.
  - Animals used for defensive purposes. What, where, when, and how.
- Ground force command and control centers in the area.
  - Center's name
  - Center's composition
  - What is the center's function
- Local ground forces integrated into the overall national defense force.
- Peak and low periods of military activity around the area.
- Local defense posture.
- Level of local civilian support to the government, to military.
- Attitude of the local military and civilian community toward the United States, and effect of their attitude on efforts during wartime.
- Access of local civilians to military facilities. If military facilities employ civilians, how many and in what positions?
- Rules of engagement for the local ground forces.
- Kinds of clothing the local civilians wear.
- Any curfews in the area, and how they are enforced.
- Normal military and civilian working hours in the area.
- Labor unions in the area and their effect on the population.
- Local currency in the area.

APPENDIX A.1

- Level of civilian/military control by a foreign government.
- Key military/civilian leaders in the area, and where they are located.
- POW handling procedures in the local area.
- Actions of the military and civilian populace under various alert conditions.
- Holiday periods observed by local military and civilian forces.
- Resistance groups located within the area.
    - Recent activity.
    - How supported from outside the country (especially if such assistance has been from the United States)?
    - Who are leaders?
    - How contacted?

B. **Naval OOB**
- Organized national naval forces located in the area.
    - Designation
    - Location
    - Manning level
    - Morale
    - Level of training
    - Uniform
    - Current activities
    - Combat effectiveness
    - Mission and functions
    - Capabilities
    - Operational limitations

NSW INTELLIGENCE

- ○ Ships and small craft (type, number, characteristics, capabilities).
- Merchant fleet ship/small craft located in the area.
    - ○ Local merchant fleet routes
    - ○ National registers of the merchant ships
    - ○ Merchant ships coastal defense functions; what are they, and how are they conducted.
- Civilian waterborne traffic found along the coast and in the harbors (e.g., tugs, water taxis, pilot craft, fishing boats, pleasure boats).
    - ○ Type
    - ○ Maritime schedule
    - ○ Function
    - ○ Normal activity
    - ○ Location.
- Surface/subsurface maritime patrols normally operating off the coast or in the harbor. What are their patrol patterns, composition, schedules, communications capabilities, reaction times.
- How is the maritime defense force integrated into the overall national coastal defense force?
- What naval force command and control centers are located in the area?
    - ○ What centers?
    - ○ What are their functions?
    - ○ What is the composition of each center?
- Water borne defensive measures/early warning techniques put into effect during an alert period (e.g., mammals, hydrophones, submerged nets, lighting, patrols).
- Security precautions currently rehearsed by ship/small craft crews.

APPENDIX A.1

- Coastal/harbor anchorages. What waterborne forces employ these anchorages?
- Is fishing conducted in the waterways at night? Local fishing regulations.
- Peak and slow times for civilian waterborne traffic in the area.
- Dredging operations being conducted, where, how, and to what depths.
- Piers located in the harbor; what forces tie up at these piers?
- Naval forces with sonar; do the forces operate their sonar in port?
- Ability of the government to draft civilian craft into military service.
- Scuba diving operations conducted in the area, by whom, where, and when?
- Waterborne rules of the road.
- Do the civilian ship crews carry weapons? If so, how many and what type?
- Is there a local UDT/EOD force? Where, what unit, how many, and what is their mission.
- Smuggling operations conducted on the waterways.
- Waterborne search and seizure regulations.
- Would naval ships deploy picket boats in the event of an alert?
- How long does it take to get the various craft underway?
- Where are the ship repair and replenishment facilities?
- Craft speed limit in the harbor.
- Local navy and civilian uniforms.
- Rubber boats operated in the area, who operates them, what type, and for what purpose?

NSW INTELLIGENCE

- Lights visible in the harbor at night.
- Does the local military have a swimmer delivery vehicle (SDV) capability?
- Units in the area which have a swimmer defense mission or swimmer detection sonar and anti-swimmer weapons.

C. Air OOB

- Organized national air forces located in the area.
    - Designation
    - Location
    - Manning level
    - Morale
    - Level of training
    - Uniform
    - Current activities
    - Combat effectiveness
    - Mission and functions
    - Capabilities
    - Operational limitations
    - Aircraft assigned and markings.
- Military and civilian airfields located in the area. What are their functions, runway characteristics, capacity, operating schedules, air routes, support, etc.
- Organized national paramilitary air forces located in the area.
- National air force reaction capability in the area.
    - What is the reaction time?
    - What are the avenues of approach to the area?
- Air forces from other nations located in the area.

APPENDIX A.1

- Air force command and control centers in the area. Where located, function, and composition?
- How do the local air forces integrate into the overall national defense posture?
- Air force air patrols in the area. What is their size, composition, operating schedule?
- Rules of engagement for air forces in the event of hostilities.
- Air navigational aids in the area, location, function, and how operated.
- Local civil air defense force.
- All weather flying capabilities of local military and civilian aircraft.
- Aircraft alert conditions and how activated.
- Air force search and rescue capability.

D. Communications OOB

- Fixed communications sites in the area.
    - Type (e.g., SATCOM, TV, radio, telephone, telegraph, etc.).
    - Site component systems.
    - Location.
    - Who operates them?
    - Technical parameters (e.g., frequencies, power outputs, ranges, modulation, types, etc.).
    - Site construction.
    - Function of each site.
    - Power supply location for each site.
    - Who is communicating on the facilities, and with whom?
    - Local call signs.

NSW INTELLIGENCE

- ○ Antennae associated with each site, and where located.
- ○ Role of each fixed site in the overall national defense posture.
- Portable communications devices in the area.
- Capability to introduce portable communication equipment into the area, what type and what reaction time.
- How can the various items of communications equipment be jammed?
- Level of compatibility between U.S. equipment and equipment in the area.
- Are there civilian ham radio operators in the area, could they be introduced into the military defense organization?
- Man-portable communications equipment in the area.
- Interface between civilian and military communications facilities in the area.
- Fixed communication cables/lines in the area, specific sites connected to each line.
- Reliance on other than electronic communications systems.

E. Electronics OOB

- Local electronics countermeasure (EM) capability.
- Local electronic counter-countermeasures (ECCM) capability.
- Local electronic support measures (EM) capability, and site location.
- Vulnerabilities of the various ECM, ECCM, ESM equipment.
- How can the various items of equipment be jammed?

APPENDIX A.1

- Power sources for the various items of equipment.
- Enemy night surveillance capability, and component location.
- Local backup capability for the various ECM, ECCM, ESM sites.
- How do these sites interface into the entire coastal defense structure?
- Enemy direction finding (DF) capability, location of sites and what are the vulnerable nodes.
- Local civilian electronic OOB, and the civilian capability to interface with the military OOB.
- What forces man the various sites, and their level of effectiveness?
- Mobile electronic OOB systems that can be introduced into the area, and their origin. Avenues of approach for such reaction force equipment.
- Undersea electronic COB in the area, and its capability to detect small craft, rubber boats (CRRC), and SDVs. What are the vulnerable nodes?
- Ground sensors in the area.

F. **Weapons OOB.**
- Missile/AAA.
    - Fixed missile sites/AAA in the area. Where are they located?
    - Function of each site.
    - When were the sites constructed?
    - Operational characteristics of the missiles.
    - Where and how are the missiles stored. How many are in storage?
    - Lowest altitude targets the surface-to-air missiles can engage.

NSW INTELLIGENCE

- ○ Launch sequence for each system, and the component parts of the launch system.
- ○ What personnel man each site (military/civilian)?
- ○ Weapon trans-shipment methods between storage and launcher.
- ○ Are the weapons in the sites pre-sighted and pre-armed?
- ○ Level of operator proficiency at each site.
- ○ Who has launch authority for the site(s)?
- ○ Vulnerable nodes for each site system.
- ○ Can the weapons in the sites be detonated sympathetically?
- ○ Minimum and maximum ranges for the weapons in each site.
- ○ Maintenance schedule for each launcher in each site.
- ○ What reload capability for each weapon in each site?
- ○ Self-propelled missiles/AAA employed in the area, what type?

- Nuclear, biological, chemical (NBC).
    - ○ Capabilities of area military to employ NBC weapons.
    - ○ NBC weapons located in the area.
    - ○ Local method of defense using such weapons.
    - ○ Local capabilities to defend against an NBC attack.
    - ○ Do troops carry and receive training with gas masks?
    - ○ Do troops train with actual NBC weapons?

APPENDIX A.1

- ○ Under what conditions would such weapons be employed?
- ○ What are the indications of the use of such weapons? (i.e., what activities would precede employment of such weapons?)
- ○ Are locals aware of the military's ability to store and employ such weapons?
* Miscellaneous.
  - ○ Types of personal weapons (e.g., rifles, machine guns, pistols, etc.) that are located/carried in the area? Who carries them?
  - ○ Do local civilians have weapons in their homes?
  - ○ Location of any armories in the area.
  - ○ Type, availability and amount of small arms ammunition in the area.
  - ○ What are the local gun control laws?
  - ○ Are the coastline, harbors, or other waterways mined? Where and what type mines?
  - ○ Are the land areas mined? Where and what type mines?
  - ○ What are the sensitivities of mines in the area and how activated?
  - ○ Do the naval/air forces have and deploy depth charges? How would they be employed?
  - ○ Does the enemy employ booby traps; how, when, where, and what type?
  - ○ Do the sentries/patrols exercise fire discipline?
  - ○ Proficiency level of the military personnel with weapons.

A.1.3 SERE

A. Evasion/Escape Routes
- Locations of nearest friendly forces
- Locations of nearest friendly borders
- Locations of nearest Safe Areas for Evacuation (SAFE)
- Cover and concealment between the target area and exfiltration points
- Water and food along exfiltration route
- Topography, vegetation, weather conditions, and dangerous wildlife, marine life, or plant life along exfiltration route
- Danger area(s) to be avoided.

B. **Pertinent Cultural Considerations.**
(See Geopolitical Brief)
- Language
- Social
- Ethnic
- Religious
- Political
- Economic
- Existence of friendly/guerrilla/underground forces/agents in the SERE area.

C. **Enemy Counter-Evasion/Escape Tactics**
Population control measures.

D. **Contact Plans**
- Location and direction (LOAD) markers
- Bonafides
- Recognition signals

- Specific locations.

A.1.4 MISCELLANEOUS INFORMATION

A. Diseases prevalent in the area, and how transmitted.

B. Camouflage measures bring employed by the military forces in the area.

C. Dispersion measures being employed by the local military forces.

D. Flags, banners, pennants that may be seen in the local area; what do they signify?

APPENDIX A.2

TARGET DEPENDENT EEI

A.2.1 IMAGERY AND GRAPHICS

A. **Area Orientation Imagery**
- Area coverage to 10 square miles (mosaic for additional area coverage required) (vertical) (10x12 format)
- Annotations:
    - North arrow
    - Installation outline
    - Other installations within imagery confines
    - Key terrain features/obstacles
    - Scale.

B. **Target Orientation Imagery**
- Area coverage to 3 square miles (vertical) (10x12 format)
- Annotations:
    - North arrow
    - Installation outline
    - Functional areas
    - Scale.

C. **Target Imagery**
- Scale - 1:5,000 (low oblique) (10x12 format)
- Annotations:
    - North arrow

APPENDIX A.2

- ○ Fences
- ○ Towers
- ○ Buildings (identify function)
- ○ All other structures/facilities (tanks, transformers, open storage, revetments, etc.)
- ○ Scale.
- Hand held photo imagery (scaled).

D. Target Graphics
- Engineering line drawing or installation blueprint
- Scale - best possible
- Keyed textual description required
- Scale and key.

A.2.2 TEXTUAL DATE/SUPPORT MATERIALS.

A. Target Description
- Physical layout/functional organization (barracks areas, maintenance areas, administrative areas, etc.)
    - ○ Number of structures/areas
    - ○ Construction of key components
        - (1) Dimensions
        - (2) Materials
        - (3) Entry/access points and type (door, ramp, loading dock, etc.)
- Primary/alternate power sources
    - ○ Number
    - ○ Type
    - ○ Location
    - ○ Conduits: location and type
    - ○ Associated facilities (transformers, switch yards, relays, etc.)

TARGET DEPENDENT EEI

- ○ Fuel supply (type and location).
- Communications associated with target:
  - ○ Type
  - ○ Number
  - ○ Location
  - ○ Associated facilities (link sites, switch center, etc.).
- On site security:
  - ○ Type (fence, ditch, passive/active detection, patrol route, etc.)
  - ○ Location
  - ○ Description:
    - (1) Dimensions
    - (2) Power source and location
    - (3) Frequency/schedule (for patrols/guards)
    - (4) Frequency/spectrum (electromagnetic).
    - (d) Internal procedures (key, cipher, personnel recognition, etc.).
- Target vulnerabilities/critical damage points:
  - ○ Type
  - ○ Location
  - ○ Dimensions
  - ○ Construction materials
  - ○ Stress point(s).
- Associated military facilities:
  - ○ Location (coordinates)
  - ○ Type force (garrison, SAM, AA, artillery, paramilitary, etc.)
  - ○ Access routes:

APPENDIX A.2

      (1) Location (from target)

      (2) Type (road, rail, waterway, etc.)

      (3) Transit time with associated transportation.

- Strength (personnel)
- Weapons (type and number)
- Organic and available transport (type and number)
- Communications with target:

      (1) Type and number

      (2) Frequency

      (3) Location of links/conduits

      (4) Alternate means of communications.

- Fuel supply:

      (1) Type fuel

      (2) Location

      (3) Access (hydrant, hose, hand pump, etc.)

      (4) Storage (tank, underground, barrel, etc.).

B. **Target Area Activity.** Include any noteworthy military or civilian activity recently associated with or in the vicinity of the target (i.e., nearby construction, observed patterns of activity).

C. **Enemy Reaction Capability.** Include a description of forces that have a capability to reinforce target security elements within short periods of time.

APPENDIX A.3

TARGET ANALYSIS CHECKLIST

A.3.1 ADMINISTRATIVE DATA

A **NAME OF FACILITY**

B **LOCATION** (ADDRESS)
- Map Coordinates
- Geographical Area
    - Urban
    - Suburban
    - Rural

C. **DATE OF ANALYSIS**

D. **AUTHOR AND SOURCES**

E. **LIST OF ATTACHMENTS**
- Maps
- Photos
- Brochures
- Schedules
- Sketches
- Blueprints.

A.3.2. GENERAL

A. **GENERAL DESCRIPTION OF FACILITY AND BRIEF COMMENTS ON NATURE OF OPERATION**

B. DESCRIPTION OF FACILITY'S COMPONENT PARTS

- Physical structure - sketch photo (air/ground) - (dimensions)
- Communications
    - Type
    - Backup systems
    - Command/control center
- Power/fuel
    - Type(s) used - (primary) - (secondary) - (alternate)
    - Amount used
        (1) daily rate
        (2) seasonal variation
    - Sources of supply
        (1) on-site storage
        (2) means of delivery and time required for resupply
    - Type of storage facility
        (1) above ground
        (2) underground
        (3) combination
    - Amount/type of fuel on hand
    - Reserve system and conversion time
        (1) type(s) used
        (2) amount used (daily rate)
        (3) sources of supply
            (a) on-site storage
            (b) means of delivery and time required for supply
        (4) type of storage facility

TARGET ANALYSIS CHECKLIST

        (a) above ground
        (b) underground
        (e) combination

- Personnel
    - Number of employees
    - Number present during each shift
    - Work hours/days
    - Key personnel (availability)
    - Labor organizations and labor/management relationships
    - Employment procedure/sharing policies.
- Raw Materials
    - Type
    - Amount
        (1) daily/weekly/monthly
        (2) stockpiles
    - Sources of supply
    - Means of delivery
- Finished product
    - Type (flammable or not)
    - Amount (daily/weekly/monthly production)
    - Quality control
    - By-products
        (1) type
        (2) amount
    - Distribution
    - Stockpile
- Conversion to manufacture of war materials
- Transportation and materials handling equipment

- Type
- Amount
- Backup system
- Maintenance/repair
- Flow diagram
- Security
  - Types of system(s)
    (1) on-site
    (2) reserve systems and reaction
  - Amount employed and schedules
  - Type of armament and how employed
  - Location
  - Screening system(s)
  - Communication systems
  - Crisis control equipment/personnel
    (1) type
    (2) amount
    (3) location
    (4) reaction time
    (5) emergency access
    (6) alarm systems
    (7) medical facilities

A.3.3. SPECIFIC

A. **LIST OF POTENTIAL TARGETS WITHIN COMPLEX**

B. **LIST OF COMMON TARGETS WITHIN COMPLEX**

TARGET ANALYSIS CHECKLIST

C. RELATIONSHIP OF TARGET TO RELATED FACILITIES/SYSTEMS SUPPORTING/DEPENDENT

1. Internal

2. External

A.3.4 CONCLUSIONS

A. **BASED UPON ANALYSIS OF TARGET COMPLEX, IDENTIFY AND JUSTIFY THOSE COMPONENTS DEEMED MOST SUSCEPTIBLE TO ATTACK BY:**
- A small force (1-12 men) with conventional weapons and explosives
- A large force (50+ men) with conventional weapons and explosives

B. **DETERMINE CONSEQUENT DOWNTIME OR DESTRUCTIVE EFFECT SUCH AN ATTACK WOULD HAVE AGAINST THE TARGET FACILITY**

APPENDIX B

VESSEL CHARACTERISTICS AND CAPABILITIES

B.1   SMALL CRAFT OPERATIONS

B.2   CRRC/INFLATABLES

B.3   SPECIAL BOAT CHARACTERISTICS

B.4   FLEET BOAT CHARACTERISTICS

B.5   SHIPS CAPABLE OF TRANSPORTING THE SEAFOX (SWCL)

B.6   NAVAL GUNFIRE SUPPORT SHIPS

APPENDIX B.1

SMALL CRAFT OPERATIONS

B.1.1 GENERAL.

The senior Unrestricted Line Officer on board any Navy craft or boat is responsible for the safe operation of the craft.

B.1.2 PLANNING CONSIDERATIONS

A. **MISSION OBJECTIVE**

B. THREAT

C. SEA/WEATHER CONDITIONS

D. LOGISTIC SUPPORT

E. EMERGENCIES

B.1.3 COORDINATION.

Boat crews should be brought into the SEAL mission planning as early as possible in order to ensure that the boat's capabilities for supporting the mission are fully exploited while also taking into consideration the boat's limitations. The planning of transits to and from the insert/extraction locations should be done jointly by the SEAL Element Commander and the boat crew to ensure that boat tactics will complement the execution of the SEAL mission.

APPENDIX B.1

B.1.4 EXECUTION.

The following items should be covered in all small craft plans:

A. PLACEMENT OF PERSONNEL AND EQUIPMENT

B. BATTLE STATIONS AND ACTIONS DURING:
- Transit
- Insertion (including deception plan)
- Extraction (including deception plan).

C. PRIMARY/SECONDARY INSERTION POINTS

D. CRAFT ACTION BETWEEN INSERTION AND EXTRACTION

E. COMMUNICATIONS/LOST COMMUNICATIONS PLAN

F. PRIMARY/SECONDARY INSERTION

G. RENDEZVOUS PROCEDURES

H. EMERGENCY EXTRACTION

I. FIRE SUPPORT

APPENDIX B.2

COMBAT RUBBER RAIDING CRAFT (CRRC)/INFLATABLES

ITEM	WEIGHT (LOBS)	LENGTH (FT)
AVON 450	270	15
AVON 460	270	15
AVON 520	350	17
Z-BIRD	400	15 (good in rough seas)
IBS	120	13
ZODIAC F-470	280	15 (preferred over AVON 460)
RIB	4,500	24
55 hp OBM	202	N/A
35 hp OBM	118	N/A
15 hp OBM	78	N/A

GAS BLADDER - 18 gal
    (13.5 gal MAX ALLOWED ON AIRCRAFT)

GAS CAN - 6 gal
    (4.5 gal MAX ALLOWED ON AIRCRAFT)

APPENDIX B.2

NOTE: Fuel consumption depends on the following variables:

- type of boat
- speed maintained
- displacement and weight of personnel and cargo
- type of motor and propeller
- engine throttle setting
- wind speed and direction
- current, set, and drift
- sea state

APPENDIX B.3

SPECIAL BOAT CHARACTERISTICS

	Mk III PB	**SWCL**	**PBR**	**MATC**
LENGTH	64'10-3/4"	35'1 1-5/8"	31'1 1"	36'
BEAM	18'3/4"	9'10"	11'7"	12'9"
HEIGHT	17'6"	6'11-3/4" COLLAPSED	8'3"	5'11"
HOISTING WEIGHT (LBS)	83,000	23,700	17,443	25,600
DISPLACEMENT (LBS)				
(LIGHT)	63,000	21,200	15,050	22,000
(FULL)	82,500	26,000	17,800	29,500
DRAFT	5'10"	2'10"	1'11-3/8" (FULL LOAD)	1'11-3/4" (FULL LOAD)
SPEED (KTS)	30+	30+	23.9	28.5
COMBAT RADIUS (FULL SPEED) (NM)	300	110	150	370
COMBAT RADIUS (REDUCED SPEED) (NM)	400	150	312	521
FUEL TYPE	#2 DIESEL	#2 DIESEL	DFM	DFM
HULL	ALUM	FBG	FBG	ALUM

APPENDIX B.3

	Mk III PB	**SWCL**	**PBR**	**MATC**
WEAPONS				
LMG	M-60	M-60	M2-56M	M-60
MG (CAL)	20MM	.50	M2-50	.50
MG	M-19	M-19	MK-46M	
MORTARS	Various			60MM
CANNON	40MM			
MAX SEA STATE	5	3	<3	<3
CREW	8	3	4	2
PAX	15 MAX	10	6	8
PAYLOAD	+CREW	SWIMMERS	+ CREW	
(LBS)	+	500		4400
COMMUNICATIONS	VRC 94	VRC 94	2	2
	ARC 159	ARC 159	UHF	UHF
	MARINE BAND			

APPENDIX B.4

FLEET BOAT CHARACTERISTICS

	LCU	LCM8	LCM6	LCPL (ST)	LCPL (AC)
LENGTH	134'9"	73'6"	56'1"	35.8'	36'
BEAM	29'9"	21'	14'	11.2'	13.5'
HEIGHT (MAST UP)	38'-1/2"				
(MAST STOWED)	19'	24'	19'	9'	9'
HOISTING WEIGHT (TONS)	200 LT	73 LT	27	9.3	8.8
DISPLACEMENT (MAX LOAD) (TONS)	390	127	62	10	10
(LIGHT LOAD)	170				
DRAFT (EMPTY/FULL)					
FWD	2.5'/4'	3.8'	3'	2.6'	3.9'
AFT	4.5'/6.5'	5.2'	4'	3.6'	
SPEED (KTS)	12	12	9	19	10
FUEL CAP (GAL)	3288	1146	466	160	160
FUEL TYPE	DFM	DFM	DFM	DFM	DFM
HULL	STEEL	STEEL	STEEL	STEEL	ALUM

APPENDIX B.4

	LCU	LCM8	LCM6	LCPL (ST)	LCPL (AC)
WEAPONS (CAL)	2-.50	2-.50	2-.50	2-.50	2-.50
CREW	10	5	5	3	3
PAX	400	150	80	17	17
COMMUNICATIONS MOTOROLA/ B TO B	Y WRC 1-B VRC 46 URC 9	Y VRD 46	Y VAL 46	Y PRC 77	Y PRC 77

APPENDIX B.5

SHIPS CAPABLE OF TRANSPORTING THE SEAFOX

SHIP TYPE	STOWAGE LOCATION/METHOD OF HANDLING
LHA	Stowage on trailer or deployment cradle (supplied by NSW units) in hangar deck, heavy vehicle deck, or forward well deck area.
LPH	Stowage in upper cradle of double banked dolly stowage. Handling by a single point lift using ship's crane.
LST	Stowage in main deck upper cradle stowage (port and/or starboard side). Handling by double pivoted link davits.
LCC	Stowage in third deck single banked cradle stowage (port and/or starboard side). Handling by trackway davits.
LSD	(28 Class): Stowage in main deck upper cradle stowage (port and/or starboard side). Handling by trackway davits. (36 Class): Stowage by nesting in LCM 6 on 01 level port side. Handling by ship's cranes. (41 Class): Stowed forward of 60 ton crane amidships on boat deck in 50 ft. utility boat cradle. Handling by 60 ton crane.
LPD	(1 Class): Stowage in 01 level upper cradle stowage (port and/or starboard side). Handling by ship's starboard crane.

APPENDIX B.5

SHIP TYPE	STOWAGE LOCATION/METHOD OF HANDLING
LPD (con't.)	(4 Class): Stowage in 01 level upper cradle stowage (port and/or starboard side). Handling by ship's starboard crane.
LKA	Stowage by nesting in LCM 8 in forward port and/or starboard locations, or on the No. 2 hatch cover. Handling by one of the 12 ship's booms.
ALL SHIPS	Additional stowage on trailer or deployment cradle (supplied by NSW units) is feasible on a flight deck or other stowage area.

NOTE: Not all ships in the classes listed have cradles configured to handle the SEAFOX. In general, the SEAFOX is capable of being stowed in any davits cradle equipped with 26,000 lb, 36-foot davits where LCPLs or LCVPs are presently stowed.

Information regarding necessary modifications to chocks, keel blocks, and gripes on LCPL and LCVP cradles to stow SEAFOX is available from the Naval Sea Systems Command, Deck and Replenishment Systems Division, Washington, D.C., 20362.

Most U.S. Navy amphibious assault ships equipped with 26,000 lb, 36-foot davits are being outfitted with a convertible cradle capable of stowing either a LCPL Mk 12 or SEAFOX. These cradles are outfitted with two sets of chocks, keel blocks and gripes (one set for each craft).

APPENDIX B.6

NAVAL GUNFIRE SUPPORT SHIPS

SHIP CLASS	NO. GUNS	RANGE (YDS)
"IOWA" BB	(9) 16 IN	40,185
	(6) TWIN 5 IN/38	17,306
"VIRGINIA" CGN	(2) 5 IN/54	25,909
"CALIFORNIA" CGN	(2) 5 IN/54	25,909
"TRUXTUN" CGN	(1) 5 IN/54	25,909
"LONG BEACH" CGN	(1) TWIN 5 IN/38	17,306
"TICONDEROGA" CGN	(2) 5 IN/54	25,909
"BELKNAP" CG	(1) 5 IN/54	25,909
"ARLEIGH BURKE" DDG	(1) 5 IN/54	25,909
"KIDD" DDG	(2) 5 IN/54	25,909
"COONTZ" DDG	(1) 5 IN/54	25,909
"CHARLES F. ADAMS" DDG	(2) 5 IN/54	25,909
"SPRUANCE" DD[1]	(2) 5 IN/54	25,909

APPENDIX B.6

SHIP CLASS	NO. GUNS	RANGE (YDS)
"OLIVER HAZARD PERRY" FFG	(1) 76MM/MK 75 (OTO MELARA)	16,300(m)
"BROOKE" FFG	(1) 5 IN/38	17,306
"KNOX FF	(1) 5 IN/54	25,909
"GARCIA" FF	(2) 5 IN/38	17,306
"BRONSTEIN" FF	(1) TWIN 3 IN/50 CAL	14,041
"TARAWA" LHA[2]	(3) 5 IN/54	25,909

The following amphibious ships have the 3 in/50 cal. twin mount which can be used in a limited NGFS role: "IWO JIMA" class LPHs, "AUSTIN" and "RALEIGH" class LPDs, "ANCHORAGE" and "THOMASTON" class LSDs, "CHARLESTON" class LKAs, and "NEWPORT" class LSTs.

---

[1] considered to be the best NGFS platform in USN due to variant of 5 in/54 gun (Mk 45) and gunfire control system (Mk 86).

[2] with the exception of "IOWA" class BBs, LHAs are the most heavily armed ship with respect to gunfire capabilities.

APPENDIX C

AIRCRAFT CHARACTERISTICS

C.1 FIXED WING AIRCRAFT

C.2 ROTARY WING AIRCRAFT

APPENDIX C.1

FIXED WING AIRCRAFT CHARACTERISTICS AND CAPABILITIES

AIRCRAFT	COMBAT RADIUS (NM)	SPEED NORM/MAX (KTS)	LANDING RQMNTS (FT)	# PAX	CARGO CAP/ PALLETS
C-1A TRADER	600	170/280	3000 X75*	10	3000
C-2 (COD)	550	260/310	1428*	28	10000
C-5A GALAXY	5600	450/496	3600 TAKE OFF: 8400	250	221000 /36Ps
C-9B SKYTRAIN	2538	438/500	5000 X75	65	32000
C-123 PROVIDER	1500	140/240	6000 X200	42	19500
C-130 HERCULES	4460	300/335	2750 TAKE OFF: 5160	92	44000 /6 Ps

FIXED WING AIRCRAFT CHARACTERISTICS

AIRCRAFT	COMBAT RADIUS (NM)	SPEED NORM/MAX (KTS)	LANDING RQMNTS (FT)	# PAX	CARGO CAP/ PALLETS
MC-130E COMBAT TALON	2000	300	CLASSIFIED		

MISC: SPECIFICALLY DESIGNED FOR A WIDE RANGE OF SPECOPS (LOW LEVEL, ALL WEATHER, NIGHT CAPABLE, INERTIAL NAV, ETC.)

AIRCRAFT	COMBAT RADIUS (NM)	SPEED NORM/MAX (KTS)	LANDING RQMNTS (FT)	# PAX	CARGO CAP/ PALLETS
AC-130H SPECTRE	CLASSIFIED				

MISC: MISSIONS INCLUDE PINPOINT FIRE SUPPORT AND TACTICAL SURVEILLANCE

AIRCRAFT	COMBAT RADIUS (NM)	SPEED NORM/MAX (KTS)	LANDING RQMNTS (FT)	# PAX	CARGO CAP/ PALLETS
C-141B STARLIFTER	5500	450	6500 X150	150	69000 /13Ps
E-2C HAWKEYE	1400	225/275	CARRIER AWACS	5	N/A
OV-10 BRONCO	265	200/341	800 X600	5	800
P-3C ORION	4000	330/350	6000 X200	23	47000

*Carrier Capable

APPENDIX C.2

ROTARY WING AIRCRAFT CHARACTERISTICS AND CAPABILITIES

AIRCRAFT	COMBAT RADIUS (NM)	SPEED NORM/MAX (KTS)	LANDING RQMNTS (FT)	PAX/ CARGO (#/LBS)
UH-1N IROQUOIS	227	100/115	65X65	11/5000
UH-1K IROQUOIS	260	110/130	65X65	11/2500
CH-46E SEA KNIGHT	180	120/145	200X100	23/6000
CH-53D SEA STALLION	223	150/170	100X100	37/8520
CH-53E SUPER SEA STALLION	226**	150/180	100X100	55/30000
HH-53H** PAVE LOW	600	170	100X100	20/30000

    MISC: SPECOPS CAPABLE, NIGHT CAPABLE, INERTIAL NAV, FLIR, ALL WEATHER, ETC.

AIRCRAFT	COMBAT RADIUS (NM)	SPEED NORM/MAX (KTS)	LANDING RQMNTS (FT)	PAX/ CARGO (#/LBS)
CH-47 CHINOOK	275	139/164	300X150	45/11650

ROTARY WING AIRCRAFT CHARACTERISTICS

AIRCRAFT	COMBAT RADIUS (NM)	SPEED NORM/MAX (KTS)	LANDING RQMNTS (FT)	PAX/ CARGO (#/LBS)
SH-3	350	120/144	100X100	6/6000
SH-60 SEAHAWK	370	145/158	100X100	N/A/6500
MH-60K**	300	146/160	100X100	15/6500
AH-6		CLASSIFIED		
AH-1T COBRA	125	150/175	65X65	not avail.

*Combat radius figures are for a normal loadout without external tanks; tradeoffs in cargo/pax capabilities will be scenario-dependent with the addition of tanks.

**In-flight refuelable.

APPENDIX D

WEAPONS AND DEMOLITIONS

D.1   **U.S. AND ALLIED SMALL ARMS**

D.2   **SOVIET/WARSAW PACT SMALL ARMS**

D.3   **DEMOLITION CAPABILITIES AND FORMULAS**

APPENDIX D.1

U.S. SMALL ARMS

WEAPON	MAG/ CAL	WT (LBS)	MAX EFF. RANGE(YDS)
MACHINE GUNS			
M-60	100 RD/ 7.62MM	23	1100
M-60 LIGHT	100 RD/ 7.62MM	19	1100
.50 BMG	100 RD/ .50 CAL	65	2200
SUBMACHINE GUNS			
MP-5	30 RD/ 9MM	4.5	110
CARBINES			
M-16 CAR	30 RD/ 5.56MM	6	440
ASSAULT RIFLES			
M-14	20 RD/ 7.62MM	11	660
M-16 A1	30 RD/ 5.56MM	7	440

APPENDIX D.1

WEAPON	MAG/CAL	WT (LBS)	MAX EFF. RANGE(YDS)
SNIPER RIFLES			
MCMILLIAN/REMINGTON M-700	5 RD/ 7.62MM	9	1000
MCMILLIAN M-86	5 RD/ 7.62MM	10	1000
.50 BMG SASR	SINGLE/ .50 CAL	30	1950
PISTOLS			
H&K P-9S	10 RD/ 9MM	2	55
.45 1911A1	8 RD/ .45 CAL	2.5	55
BERETTA 92F	15 RD/ 9MM	2	55
S&W MOD 686	6 RD/ .357 CAL	2.5	55
MISCELLANEOUS			
REMINGTON MOD 870	SHOTGUN/ 12 GUAGE		55
M-203 GRENADE LAUNCHER	Single Shot/ 40MM	3	330
M-3 CARL-GUSTAF RECOILESS RIFLE	Single Shot/ 84MM	19	1100
SHOULDER MOUNTED ANTI-ARMOR WEAPON (SMAW)		16	330
AT-4 ANTI-ARMOR	Single Shot/ 84MM	12	330

PATROL LEADER'S HANDBOOK

APPENDIX D.2

SOVIET/WARSAW PACT SMALL ARMS

WEAPON	CAL	WT (LBS)	MAX EFF. RANGE(YDS)
MACHINE GUNS			
PKM (BELT FED)	7.62MM	44	1100
RPK (MAG. FED)	7.62MM	28	875
ASSAULT RIFLES			
AK-47	7.62MM	9.5	330
AKM OR AKMS	7.62MM	7	330
AK-74/AKS-74	6.64MM	8	
SVD SNIPER RIFLE	7.62MM	9	660
PISTOLS			
MAKAROV PM	9MM	1.4	55
PSM	5.54MM	1	55
ANTI-ARMOR			
RPG-7	40MM	17	330
RPG-18	64MM	6	220

APPENDIX D.3

DEMOLITIONS CAPABILITIES AND FORMULAS

DEMOLITIONS CAPABILITIES

DEMOLITION CARD   GTA 5-10-9

SUPERSEDES GTA 5-10-9, MAY 1965   DECEMBER 1969

DISTRIBUTION: Active Army, ARNG, USAR: To be distributed in accordance with DA Form 12-12. See II requirements pertinent to TOE, 5, 7, 17, and 31 series

SEE AR 385-63, SAFETY   ☆ GPO : 1970 O - 370-138

	RELATIVE EFFECTIVENESS OF MILITARY EXPLOSIVES USED AS EXTERNAL CHARGES		SIZE ISSUED
1.00	TNT	(BLOCK)	1/2 LB, 1 LB
1.34	COMPOSITION C-4	(M 5A1 DEMO BLOCK)	2 1/2 LB
	COMPOSITION C-4	(M112 DEMO BLOCK)	1 1/4 LB
1.20	TETRYTOL	(M-1 AND M-2 DEMO BLOCK)	2 1/2 LB
0.42	CRATERING CHARGE	(AMMONIUM NITRATE)	40 LB
0.92	MILITARY DYNAMITE	(M-1)	1/2 LB
1.14	SHEET EXPLOSIVE	(M118)	1/2 LB
1.14	SHEET EXPLOSIVE (ROLL)	(M186)	25 LB

Quantities of explosives in these formulas and tables are for TNT, for other explosives, divide the quantity for TNT by the effectiveness factor.

MINIMUM SAFE DISTANCES FOR PERSONS IN THE OPEN WITH BARE CHARGES

POUNDS OF EXPLOSIVE	SAFE DISTANCE IN FEET	POUNDS OF EXPLOSIVE	SAFE DISTANCE IN FEET
1 TO 27 INCL	900	150	1583
32	961	200	1752
40	1020	300	2007
50	1104	400	2208
80	1290	500	2382
100	1392	OVER 500	d = 300 ∛lbs of exp

MINIMUM SAFE DISTANCE FOR PERSONS IN MISSILE PROOF SHELTER IS 300 FT.

SAFETY REMINDERS, DO'S & DONT'S

1. DO NOT HANDLE EXPLOSIVES CARELESSLY
2. DO NOT DIVIDE RESPONSIBILITY FOR EXPLOSIVE WORK
3. DO NOT MIX EXPLOSIVES AND DETONATORS
4. DO NOT CARRY EXPLOSIVES OR CAPS IN POCKETS
5. DO KEEP BLASTING MACHINE UNDER CONTROL OF NCOIC.
6. DO WEAR HELMETS AT ALL TIMES WHILE FIRING EXPLOSIVES.
7. DO HANDLE MISFIRES WITH EXTREME CARE
8. DO NOT TAKE CHANCES

PROBLEM SOLVING FORMAT

1. CALCULATE NUMBER OF POUNDS OF TNT REQUIRED
2. DIVIDE BY RELATIVE EFFECTIVENESS FACTOR.
3. ROUND UP TO NEXT POUND OR PACKAGE SIZE
4. SOLVE FOR NUMBER OF CHARGES
5. SOLVE FOR TOTAL EXPLOSIVES

CONVERSION FACTORS FOR ALL TABLES

1 METER = 3.28 FT
1 KILOGRAM = 2.20 LB
1 FT = .3048 METER
1 LB = .4536 KILOGRAM

APPENDIX D.3

STEEL CUTTING CHARGES
POUNDS TNT = 3/8 × AREA OF CROSS SECTION IN SQ IN
(CALCULATE RECTANGULAR AREAS, THEN ADD TO OBTAIN TOTAL AREA)

EXAMPLE PROBLEM

FLANGES: WIDTH = 8"
THICKNESS = 5/8"

CHARGE: FROM TABLE = 19

WEB: WIDTH = 18"
THICKNESS = 1"

CHARGE: FROM TABLE = 68

CHARGE TOTAL:
2 FLANGES = 2×19 = 38
WEB = 68
TOTAL = 106

USE 11 POUNDS TNT

PLACEMENT OF CHARGES ON STEEL MEMBERS

PLASTIC EXPLOSIVE ON CHANNEL | TNT PLACED ON ONE SIDE OF I BEAM | EXPLOSIVE CHARGE DIVIDED IN HALF, OFFSET MINIMUM THICKNESS OF WEB

CABLES | RODS | BARS

FOR CUTTING HIGH CARBON STEEL PARTS, ALLOY STEEL ARTICLES, OR SLENDER STEEL MEMBERS P(TNT) = D³

RULE OF THUMB FOR MILD CIRCULAR STEEL SECTIONS

LESS THAN 1 INCH USE 1 LB OF TNT
1 INCH AND OVER BUT LESS THAN 2 INCHES USE 2 LB OF TNT
2 INCHES AND OVER USE P = 3/8A

THICKNESS OF SECTION IN INCHES	POUNDS OF TNT FOR RECTANGULAR STEEL SECTIONS OF GIVEN DIMENSIONS - WIDTH OF SECTION IN INCHES											
	2	3	4	5	6	8	10	12	14	16	18	20	24
1/4	0.2	0.3	0.4	0.5	0.6	0.8	1.0	1.2	1.3	1.5	1.7	1.9	2.3
3/8	0.3	0.5	0.6	0.7	0.9	1.2	1.4	1.7	2.0	2.3	2.6	2.8	3.4
1/2	0.4	0.6	0.8	1.0	1.2	1.5	1.9	2.3	2.7	3.0	3.4	3.8	4.5
5/8	0.5	0.7	1.0	1.2	1.4	1.9	2.4	2.9	3.3	3.8	4.3	4.7	5.7
3/4	0.6	0.9	1.2	1.4	1.7	2.3	2.8	3.4	4.0	4.5	5.1	5.7	6.8
7/8	0.7	1.0	1.4	1.7	2.0	2.7	3.3	4.0	4.6	5.3	6.0	6.6	7.9
1	0.8	1.2	1.5	1.9	2.3	3.0	3.8	4.5	5.3	6.0	6.8	7.5	9.0

TO USE TABLE:
1. MEASURE RECTANGULAR SECTIONS OF MEMBER SEPARATELY
2. USING TABLE, FIND CHARGE FOR EACH SECTION
3. ADD CHARGES FOR SECTIONS TO FIND TOTAL CHARGE.
4. NEVER USE LESS THAN CALCULATED CHARGE.
5. IF DIMENSION IS NOT ON TABLE, USE NEXT LARGER DIMENSION

DEMOLITIONS CAPABILITIES

PRESSURE CHARGES
P = 3H²T (ADD 1/3 IF CHARGE IS UNTAMPED)
MINIMUM TAMPING - 10"

USED ON SIMPLE SPANS, CONCRETE T BEAM BRIDGES

PLACE CHARGES AT MIDSPAN
CONSTRUCTION JOINT
CHARGES

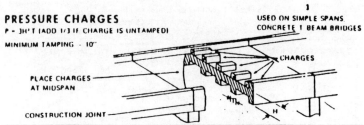

POUNDS OF TNT FOR EACH BEAM-TAMPED CHARGES

HEIGHT OF BEAM IN FEET	THICKNESS OF BEAM IN FEET								
	1 12 IN	1¼ 15 IN	1½ 18 IN	1¾ 21 IN	2 24 IN	2¼ 27 IN	2½ 30 IN	2¾ 33 IN	3 36 IN
1 (12 IN)	3								
1¼ (15 IN)	5	6							
1½ (18 IN)	7	9	11						
1¾ (21 IN)	10	12	14	16					
2 (24 IN)	12	15	18	21	24				
2¼ (27 IN)	16	19	23	27	31	35			
2½ (30 IN)	19	24	29	33	38	43	47		
2¾ (33 IN)	23	29	34	40	46	51	57	63	
3 (36 IN)	27	34	41	48	54	61	68	75	81
3¼ (39 IN)	32	40	48	56	64	72	80	88	96
3½ (42 IN)	37	46	56	65	74	83	92	101	111
3¾ (46 IN)	43	53	64	74	85	95	106	116	127
4 (48 IN)	48	60	72	84	96	108	120	132	144
4¼ (51 IN)	56	68	82	96	109	122	136	149	163
4½ (54 IN)	61	76	92	107	122	137	152	167	183
4¾ (57 IN)	68	85	102	119	136	153	170	187	203
5 (60 IN)	75	94	113	132	150	169	188	207	225

TIMBER CUTTING CHARGES

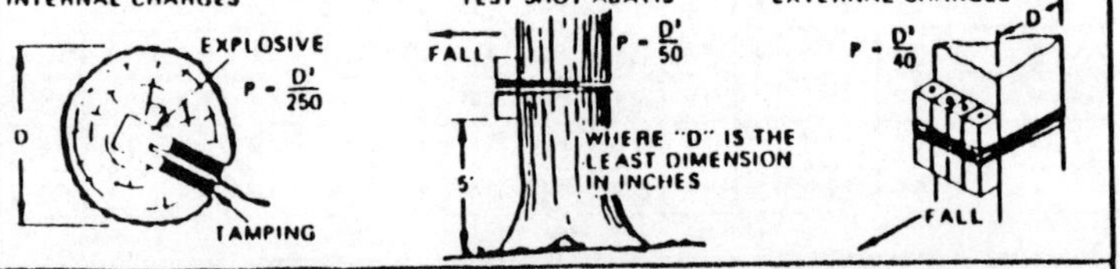

INTERNAL CHARGES — EXPLOSIVE — $P = \frac{D^2}{250}$ — TAMPING

TEST SHOT ABATIS — FALL — $P = \frac{D^2}{50}$ — WHERE "D" IS THE LEAST DIMENSION IN INCHES — 5'

EXTERNAL CHARGES — $P = \frac{D^2}{40}$ — FALL

TYPE OF CHARGE	EXPLOSIVE	LEAST DIMENSION OF TIMBER IN INCHES											
		6	8	10	12	15	18	21	24	27	30	33	36
		POUNDS OF EXPLOSIVE											
INTERNAL	ANY	¼	¼	½	1	1	1½	2	2½	3	4	4½	5¼
EXTERNAL	TNT	1	2	2½	4	6	8½	11½	14½	18¼	22½	27½	32½
ABATIS	TNT	1	1¼	2	3	4½	8½	9	11½	14½	18	22	26

APPENDIX D.3

CRATERING CHARGES

DELIBERATE ROAD CRATER

ALTERNATE 5 FT AND 7 FT HOLES SPACED ON 5 FT CENTERS

NO TWO 5 FT HOLES ARE TO BE PLACED NEXT TO EACH OTHER (END HOLES ALWAYS 7 FT)

USE 40 LB CHARGES IN 5 FT HOLES AND 80 LB CHARGES IN 7 FT HOLES

RESULTING CRATER APPROX 8 FT DEEP AND 25 FT WIDE.

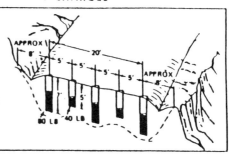

FORMULA FOR NUMBER OF HOLES $N = \frac{L-16}{5} + 1$

HASTY ROAD CRATER

HOLES OF EQUAL DEPTH (2½ FT TO 5 FT)

USE 10 POUNDS OF EXPLOSIVE PER FT OF DEPTH

RESULTING CRATER DEPTH APPROX 1½ TIMES DEPTH OF BORE HOLES

WIDTH APPROX 5 TIMES DEPTH OF BORE HOLES

NOTE: ALL CRATERING CHARGES TO BE DUAL PRIMED WITH AT LEAST ONE LB OF EXPLOSIVE

RELIEVED-FACE CRATERING

LAYOUT FRIENDLY ROW FIRST AS SHOWN. LAYOUT ENEMY ROW WITH HOLES CENTERED BETWEEN FRIENDLY HOLES. DETONATE ENEMY ROW FIRST DETONATE FRIENDLY ROW ½ TO 1½ SEC DELAY AFTER ENEMY ROW

FORMULA FOR NUMBER OF HOLES FRIENDLY ROW =

$N = \frac{L-10}{7} + 1$

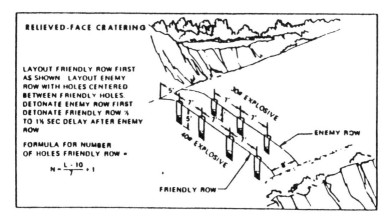

DEMOLITIONS CAPABILITIES

BREACHING CHARGES

4. USING THE TABLE BOTTOM OF P. 5, DETERMINE THE APPROPRIATE CONVERSION FACTOR
5. USING THE TABLE TOP OF PAGE 5, DETERMINE THE AMOUNT OF TNT THAT WOULD BE REQUIRED IF THE OBJECT WERE MADE OF REINFORCED CONCRETE.
6. MULTIPLY THE NUMBER OF POUNDS OF TNT (FROM TABLE) BY THE CONVERSION FACTOR

EXAMPLE:

A TIMBER AND EARTH WALL 6½ FT. THICK AND AN EXPLOSIVE CHARGE PLACED AT THE BASE OF THE WALL WITHOUT TAMPING. THE CONVERSION FACTOR IS 0.5 (SEE TABLE BOTTOM P. 5). IF THIS WALL WERE MADE OF REINFORCED CONCRETE, 618 LBS. OF TNT WOULD BE REQUIRED TO BREACH IT (SEE TABLE TOP PAGE 5). MULTIPLY 623 LBS. OF TNT BY 0.5 AND THE RESULT IS 312 LBS. OF TNT REQUIRED TO BREACH IT.

BREACHING CHARGE FORMULA

$P = R^3 KC$ where,
P = POUNDS OF TNT REQUIRED,
R = BREACHING RADIUS, IN FEET,
K = MATERIAL FACTOR (FROM TABLE PAGE 7),
C = TAMPING FACTOR (FROM TABLE BELOW).

EXAMPLE: BREACH A 4 FT. REINFORCED CONCRETE WALL WITH AN UNTAMPED ELEVATED CHARGE.

$P = R^3 KC$, $R = 4$, $K = 80$, $C = 1.8$
$P = 4^3 \times 80 \times 1.8$
$P = 9216$ LBS. TNT, USE 93 LBS.

ROUND OFF RULE FOR N

N = LESS THAN 1.25, USE 1 CHARGE.
N = 1.25 TO 2.50, USE 2 CHARGES.
N = OVER 2.50, ROUND OFF TO NEAREST WHOLE NUMBER.

NUMBER OF CHARGES = $N = \dfrac{W \text{ (width)}}{2R \text{ (breaching radius)}}$

VALUES OF C

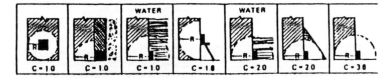

RULES OF THUMB FOR BREACHING CHARGES

FOR BEST RESULTS PLACE CHARGE IN SHAPE OF A FLAT SQUARE, FLAT SIDE TO THE TARGET.
FOR CHARGES LESS THAN 40 LBS., USE CHARGE THICKNESS OF 2" (1 BLOCK THICK).
FOR CHARGES 40 LBS. OR MORE, USE CHARGE THICKNESS OF 4" (1 HAVERSACK THICK).
FOR BREACHING HARD SURFACE PAVEMENT, USE 1 LB. FOR EACH 2" OF PAVEMENT (TAMPING = 2 × THICKNESS OF PAVEMENT)
FOR CONCRETE OBSTACLES 100 CUBIC FEET OR LESS, USE 1 LB/CU.FT. (TETRYTOL OR HIGHER).

APPENDIX D.3

BREACHING CHARGES
REINFORCED CONCRETE ONLY

(FOR OTHER TYPES OF CONSTRUCTION SEE BELOW)

THICKNESS OF CONCRETE	METHODS OF PLACEMENT					DISTANCE BETWEEN CHARGES	
						INTERNAL	EXTERNAL
FEET	POUNDS OF TNT					FEET	FEET
2	2	8	14	18	28	2	4
2½	2	15	27	30	64	2½	5
3	4	22	39	44	78	3	6
3½	6	36	62	69	124	3½	7
4	8	52	83	103	186	4	8
4½	11	73	132	146	263	4½	9
5	16	79	142	168	284	5	10
5½	20	106	189	210	378	5½	11
6	22	138	246	273	490	6	12
6½	28	173	312	346	623	6½	13
7	36	188	334	371	667	7	14
7½	43	228	410	456	821	7½	15
8	52	277	498	553	996	8	16

TABLE ABOVE IS FOR REINFORCED CONCRETE ONLY.
FOR OTHER TYPES OF MATERIAL USE THE FOLLOWING CONVERSION FACTORS.

EARTH	ORDINARY MASONRY, HARDPAN, SHALE, ORDINARY CONCRETE, ROCK, GOOD TIMBER AND EARTH CONSTRUCTION	DENSE CONCRETE, FIRST CLASS MASONRY
0.1	0.5	0.7

TO USE TABLES IN CALCULATING BREACHING CHARGES

1. DETERMINE THE TYPE OF MATERIAL IN THE OBJECT YOU PLAN TO DESTROY. IF IN DOUBT, ASSUME THE MATERIAL TO BE OF THE STRONGER TYPE – eg. UNLESS YOU KNOW DIFFERENTLY, ASSUME CONCRETE TO BE REINFORCED.
2. MEASURE THICKNESS OF OBJECT.
3. DECIDE HOW YOU WILL PLACE THE CHARGE AGAINST THE OBJECT. COMPARE YOUR METHOD OF PLACEMENT WITH THE DIAGRAMS AT THE TOP OF THE PAGE. IF THERE IS ANY QUESTION AS TO WHICH COLUMN TO USE, ALWAYS USE THE COLUMN THAT WILL GIVE YOU THE GREATER AMOUNT OF TNT.

DEMOLITIONS CAPABILITIES

BREACHING CHARGES

FORMULA P(TNT) = R^3 KC

VALUES OF K		
MATERIAL	R	K
EARTH	ALL VALUES	0.07
POOR MASONRY, SHALE, HARDPAN, GOOD TIMBER AND EARTH CONSTRUCTION	LESS THAN 5 FT 5 FT. OR MORE	0.32 0.29
GOOD MASONRY CONCRETE BLOCK ROCK	1 FT OR LESS OVER 1 FT. TO LESS THAN 3 FT. 3 FT. TO LESS THAN 5 FT. 5 FT. TO LESS THAN 7 FT. 7 FT. OR MORE	0.88 0.48 0.40 0.32 0.27
DENSE CONCRETE, FIRST CLASS MASONRY	1 FT OR LESS OVER 1 FT. TO LESS THAN 3 FT. 3 FT. TO LESS THAN 5 FT. 5 FT. TO LESS THAN 7 FT. 7 FT. OR MORE	1.14 0.62 0.52 0.41 0.36
REINFORCED CONCRETE (CONCRETE ONLY; WILL NOT CUT REINFORCING STEEL)	1 FT. OR LESS OVER 1 FT. TO LESS THAN 3 FT. 3 FT. TO LESS THAN 5 FT. 5 FT. TO LESS THAN 7 FT. 7 FT. OR MORE	1.76 0.96 0.80 0.63 0.54

BRIDGE ABUTMENT DESTRUCTION

ABUTMENTS 5 FT OR LESS IN THICKNESS

BEGINNING 5 FT IN FROM ONE SIDE OF ROAD, PLACE 40-LB CRATERING CHARGE IN HOLES 5 FT DEEP, 5 FT ON CENTERS AND 5 FT BEHIND RIVER FACE OF ABUTMENT

IF ABUTMENTS ARE OVER 20 FT HIGH, ADD A ROW OF BREACHING CHARGES ON THE RIVER FACE OF THE ABUTMENT

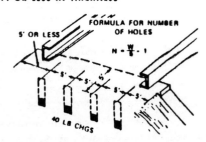

FORMULA FOR NUMBER OF HOLES

$N = \frac{W}{5} - 1$

ABUTMENTS MORE THAN 5 FT THICK

CALCULATE CHARGES BY BREACHING FORMULA AND PLACE AGAINST REAR FACE AT A DEPTH EQUAL TO THICKNESS OF ABUTMENT AND SPACE THE SAME AS OTHER BREACHING CHARGES

WHEN ABUTMENTS ARE MORE THAN 20 FT HIGH, ADD A ROW OF BREACHING CHARGES ON THE RIVER FACE AT THE BASE OF THE ABUTMENT AND FIRE ALL CHARGSE SIMULTANEOUSLY

APPENDIX D.3

ADVANCED DEMOLITION TECHNIQUES
THESE TECHNIQUES ARE INTENDED TO SUPPLEMENT CONVENTIONAL FORMULAS

DIAMOND CHARGE
LONG AXIS - CIRCUMFERENCE OF TARGET
SHORT AXIS - ½ CIRCUMFERENCE OF TARGET
THICKNESS OF CHARGE: 1 INCH
C4 BLOCK SHOULD BE CUT NOT MOLDED.
DETONATION: SIMULTANEOUS AT EACH END OF SHORT AXIS

SADDLE CHARGE
BASE - ½ CIRCUMFERENCE
LONG AXIS - 2 TIMES BASE
THICKNESS OF CHARGE: 1 INCH C4 UP TO 8" DIAM
DETONATION: FROM APEX OF TRIANGLE DIAM ONLY FOR MILD STEEL UP TO 8"

RIBBON CHARGE
DEPTH: ½ THICKNESS OF TARGET
WIDTH: 3 TIMES THICKNESS OF CHARGE
LENGTH: SAME AS LENGTH OF CUT DESIRED

COUNTER-FORCE CHARGE
SIZE: 1¼ LB PER FT OF CONCRETE. PLACE HALF OF CHARGE ON EACH SIDE OF TARGET, DIRECTLY OPPOSITE EACH OTHER. BOTH CHARGES MUST BE DETONATED SIMULTANEOUSLY. USE ON CUBES & COLUMNS, NOT WALLS OR PIERS.

SHAPED CHARGE
HEIGHT OF CHARGE - 2 x HEIGHT OF CONE
CONE ANGLE - 45° TO 60°
STANDOFF - 1½ x DIAM. OF CONE
DETONATION: EXACT TOP CENTER OF CHARGE

PLATTER CHARGE
EXPLOSIVE WT SHOULD BE APPROX EQUAL TO PLATTER WT. 2 TO 6 LB RECOMMENDED.
DETONATION: FROM EXACT REAR CENTER. PLATTER NEED NOT BE ROUND OR CONCAVE.

MATERIAL	M3			M2A3		
	PENETRATION	DIA. OF HOLE	STANDOFF	PENETRATION	DIA. OF HOLE	STANDOFF
REINFORCED CONCRETE	60"	3½"	STANDARD	30"	2½"	STANDARD
ARMOR PLATE	20"	2½"	STANDARD	12"	1½"	STANDARD
PERMAFROST	72"	6" TO 8"	60"	72"	6" TO 1½"	30"
ICE	12'	6"	42"	7'	3½"	42"
SOIL	7'	14.6"	48"	7'	7"	30"

RULES OF THUMB

RAILROAD RAILS OVER 80 LB. PER YD. (MORE THAN 5" HIGH)   1 LB. TNT
RAILROAD RAILS 80 LB. OR LESS PER YD. (5" OR LESS HIGH)   1/2 LB. TNT
DITCH, PER CU. YD. OF EARTH   1 LB.
DEAD STUMPS, PER FT. OF DIAMETER } (ADD 60% FOR   1 LB
GREEN STUMPS, PER FT. OF DIAMETER } STANDING TIMBER)   2 LB.

APPENDIX E

COMMUNICATIONS

E.1 COMMUNICATIONS/ELECTRONICS CAPABILITIES AND SPECIFICATIONS

E.2 NSW/SHIPBOARD COMMUNICATIONS INTEROPERABILITY

E.3 NSW/E-2C INTEROPERABILITY

E.4 C3 VAN CAPABILITIES

APPENDIX E.1

COMMUNICATIONS/ELECTRONICS CAPABILITIES AND SPECIFICATIONS

ITEM	FREQ RANGE (MHZ)	RANGE (NM)	WT. (LBS)
AN/PRC-68	30-79	1	3
AN/PRC-77	30-75.95	25	23
AN/PRC-90	243-282.8	LOS (EMERG)	2
URC-94	1.5-29.99/ 30.0-79.99	50	N/A
AN/PRC-104	2-29.99	10-2000	28
AN/URC-110	116-149/ 225-399	5-5000 (L.O.S.) SATCOM	16 (KY-65)
AN/URC-112	121.5-225.9	LOS REPLACING URC-90	1.5
AN/PRC-113	116-150/ 225-400	LOS	10
AN/PRC-117	30-89.975	5 REPLACING PRC-77	11
PRC-119	30-88	LOS	18
ARC-159	225-395.975	LOS	N/A
AN/PSC-3	225-400	LOS/ SATCOM	25
LST-5A	225-400	LOS/ SATCOM	8
KY-57	VINSON CRYPTO	N/A	6
KY-65	PARKHILL	N/A	25

APPENDIX E.1

ITEM	FREQ RANGE (MHZ)	RANGE (NM)	WT. (LBS)
MX-300	139.6 (VHF)	5	2
MX-300	407.425/.525	5	2
MX-340	156.8	5	2
DMDG	HIGH SPEED COMMUNICATIONS (NOT SECURE)		9
AN/PPN-18	BEACON	N/A	26
AN/PPN-19	BEACON	N/A	17
M-909	NIGHT VISION GOGGLES (NEW)	N/A	4
PVS-5A	NIGHT VISION GOGGLES (OLD)	N/A	4
M-845	NIGHT RIFLE SCOPE	N/A	4
M-911	POCKET SCOPE	N/A	2
PAQ-4	IR AIMING LIGHT	N/A	2
GVS-5	IR RANGE FINDER	N/A	6
AIM-1	IR AIMING LIGHT	N/A	1

APPENDIX E.2

NSW/SHIPBOARD COMMUNICATIONS INTEROPERABILITY

	ADVANCE FORCE CDR				CATF			BG	
	LSD	**LPD**	**LST**	**FF**	**LCC**	**LPH**	**LHA**	**CV**	**BB**
PLATOON									
PRC-104 (HF)	+	+	+	+	+	+	+	+	+
									Note 1
PRC-113 (HIGH VHF/UHF)	+	+	+	+	+	+	+	+	+
PRC-112 (UHF)	+	+	+	+	+	+	+	+	+
PRC-117 (VHF in non-hop Mode)	+	+	+	+	+	+	+	+	+
									Note 2
PRC-68 (VHF)	+	+	+	+	+	+	+	+	+
PRC-77 (VHF)	+	+	+	+	+	+	+	+	+
PRC-90 (UHF)	+	+	+	+	+	+	+	+	+
PRC-94 (UHF)	+	+	+	+	+	+	+	+	+
PSC-3 (SATCOM)	*	*	*	*	*	*	*	*	*
URC-110 (SATCOM)	*	*	*	*	*	*	*	*	*
MX300R (HIGH VHF)	-	-	-	-	-	-	-	-	-
NSWTU									
GRC-193 (HF)	+	+	+	+	+	+	+	+	+
URC-94 (HF/VHF)	+	+	+	+	+	+	+	+	+
PRC-113 (UHF)	+	+	+	+	+	+	+	+	+
VSC-7 (SATCOM)	*	*	*	*	*	*	*	*	*

APPENDIX E.2

	ADVANCE FORCE CDR				CATF			BG	
	LSD	**LPD**	**LST**	**FF**	**LCC**	**LPH**	**LHA**	**CV**	**BB**
NSWTC									
URT-23 (HF TRANS)	+	+	+	+	+	+	+	+	+
R1051 (HF REC)	+	+	+	+	+	+	+	+	+
URC-94 (HF/VHF)	+	+	+	+	+	+	+	+	+
VSC-7 (SATCOM)	*	*	*	*	*	*	*	*	*
WSC-3 (SATCOM)	*	*	*	*	*	*	*	*	*

+ Compatible with organic ship radios.
- Not compatible with ship radios.
* Compatibility with ship radios depends on MODs of ship equipment.

Note 1 - All DMDG operations (w/any radio) require NSW personnel and equipment to interface shipboard equipment.
Note 2 - Ship equipment is not compatible with frequency hopping mode.

APPENDIX E.2.1

NSW/SHIPBOARD SATCOM INTEROPERABILITY

A. Existing shipboard assets (WSC-3, OE-82, KY-58) can be utilized. NSW supplies DMDG and TA-970 adapter (WILLY BOX).

    1. WSC-3 must be solely dedicated to NSW forces.

    2. If no TA-970/TA-790 available, WILLY BOX would have to be adapted to plug directly into KY-58. This limits remoting capability.

B. Utilizing NSW gear (PSC-3, KY-57, DMDG)

    1. Deck mounting of DMC-120 portable SATCOM antenna is required. This limits area of PSC-3 due to length of antenna cable.

    2. The possibility exists to tap off existing shipboard OE-82 SATCOM antenna. Some ships are so equipped. In other cases, NSW forces would have to provide hardware.

    3. An additional option would be to strap DMC-120 antenna to OE-82 antenna for tracking purposes.

    4. In cases 2 and 3, some satellite must be used.

APPENDIX E.3

NSW/E-2C INTEROPERABILITY

1. Cable required for DMDG to E-2C.

2. UHF LOS (PRC-113), DMDG, KY-57 operational on testing.

3. HF (PRC-104), DMDG operational.

4. Aircraft altitude will effect range of communication. (25,000 ft. at 300 miles for UHF).

APPENDIX E.4

C3 VAN CAPABILITIES

C-3 VAN QTY. CIRCUIT	CIRCUIT CAPABILITY
4 HF	TTY/VOICE, AM/USB/LSB/CW/LANDLINE
2 HF/VHF	VOICE/DATA, AM/USB/LSB/CW/FM-VHF
*3(2) UHF-FM	VOICE/DATA, LOS/SATCOM
1 UHF	AM/FM, VOICE/DATA, FSK, LOS/SATCOM
4 VINSON	SECURE VOICE (KY-58) SATCOM, UHF, VHF
2 PARKHILL	VOICE (KY-75) DATA, HF, LANDLINE
2 NESTOR	SECURE VOICE (KY-8) UHF, VHF
1	FLTSEVOCOM (STANDARD FLEET SECURE VOICE NET)
*1(0) DAMA	DEMAND ASSIGNED MULTIPLE ACCESS (ALLOWS FIVE CIRCUITS TO TIME-SHARE ONE SATELLITE CHANNEL)
*1(0) GXC-7	DIGITAL FACSIMILE CIRCUIT, SECURE/NON-SECURE
2 KG-84	DIGITAL DATA ENCRYPTION DEVICES (TTY/DATA) ENABLES HIGH-SPEED INFORMATION TRANSFER
2 DMDG	DIGITAL MESSAGE DEVICE GROUP

Numbers in parentheses indicate current circuit capability. Full circuit capability will be added in the near future. All wiring is installed in the van(s), with the exception of DAMA, but the equipment itself is not yet available.

APPENDIX F

REFERENCES

F.1 **BIBLIOGRAPHY FOR NSW OPERATIONS PLANNING**

F.2 **GLOSSARY OF NAVAL SPECIAL WARFARE TERMS**

APPENDIX F.1

BIBLIOGRAPHY FOR NSW OPERATIONS PLANNING

F.1.1 AMPHIBIOUS OPERATIONS

Naval Special Warfare in Amphibious Operations	NWP 22-4B
Joint Surf Manual	COMNAVSURF-PAC/LANT Inst 3840.1 (Series)

F.1.2 ENVIRONMENTAL AREAS OF OPERATIONS

Desert Operations	FM 90-3
Mountain Operations	FM 31-72
Northern Operations	FM 31-71
Basic Cold Weather Operations	FM 31-70

APPENDIX F.1

F.1.3 CARTOGRAPHY

Nautical Chart Symbols and Abbreviations	H.O. Chart No. 1.
Joint Surf Manual	COMNAVSURF-PAC/LANT Inst 3840.1 (Series)
Cartographer's Manual	CNSWG-1 Inst 3820.1

F.1.4 DEMOLITIONS

Demolitions Materials	NAVSEA OP 2212
Explosives and Demolitions	FM 5-25
Special Forces Explosives	Army Correspondence Course Sub-Course 706

F.1.5 DIVING

Navy Diving Manual	Vol. I Air Diving Vol. II Mixed Gas Diving
Draeger LAR V	Pamphlet
Closed-Circuit 02 Diving	NEDU Rpt 7-85
U/W Purging Procedures for Draeger	NEDU Rpt 8-86

F.1.6 LAND WARFARE

The Law of Land Warfare	FM 27-10
Map Reading	FM 21-26
Ranger Handbook	ST 21-75-2
Sniper Training and Employment	TC 23-14
Special Forces Trainer's Guide	FM 31-5
Survival	FM 21-76

F.1.7 PHOTOGRAPHY

Intelligence Photography	FITCPAC Course J-243-0974

F.1.8 SOVIET/EASTERN BLOC FORCES

The Soviet Army Troops, Org., and Equipment	FM 100-2-3
The Soviet Army, Operations and Tactics	FM 100-2-1
Understanding Soviet Naval Developments	NAVSO P-356
Visual Aircraft Recognition	FM 44-30

APPENDIX F.1

F.1.9 SUBMARINE OPERATIONS

Submarine SPECOPS Manual - Unconventional Warfare	NWP 79-0-4
DDS TACMEMO	NAVSPECWAR-CEN STG TACMEMO

F.1.10 TARGETING

SPECOPS Target Vulnerability and Weaponeering Manual	61JTCG/ME-83

F.1.11 WEAPONS

M-16 A1 Rifle	FM 23-9
M-14 Rifle	FM 23-8
M-60 Machine Gun	FM 23-67
12 Gauge Shotgun	TM 9-1005-303-14
60mm Lightweight Mortar	TM 9-1010-223-10-HR
40mm M-203 Grenade Launcher	FM 23-31
Pistols and Revolvers	FM 23-35

APPENDIX F.2

GLOSSARY OF NAVAL SPECIAL WARFARE TERMS

Antiterrorism: Defensive measures used to reduce the vulnerability of individuals or property to terrorism. Also called AT (approved definition by JCS Pub 1).

Beach Landing Site (BLS): A geographical location selected for across the beach infiltration/exfiltration/resupply operations.

Beacon Bombing: Bombing operations using Radar Beacon Forward Air Controller (RABFAC) AN/PPN-18 and AN/PPN-19 transponders to aid aircraft in the conduct of close air support missions. Often used in conjunction with ground laser devices to deliver precision guided munitions.

Blind Transmission: Transmission which is without expectation of a receipt or reply.

Brown Water: An unofficial term, generally used to encompass riverine, inshore, and coastal operations. "Riverine" is an inland or coastal area, characterized by both land and water, with limited land routes and extensive water surface and/or inland waterways. "Inshore" relates to coastal areas and is generally used to indicate activities adjacent to the shore (i.e., in very shallow water). "Coastal" is the least defined term, generally taken to mean over the continental shelf (i.e., a depth of 600 ft or less).

Civil Affairs: Those activities conducted during peace and war that facilitate relationships between US military forces, civil authorities, and people of the nation in which the U.S. forces are operating.

APPENDIX F.2

<u>Clandestine Operations:</u> Operations to accomplish intelligence, counterintelligence, and other similar activities sponsored or conducted by governmental agencies in such a way as to assure concealment of identity of sponsor.

<u>Combat Control Team:</u> A team of Air Force personnel organized, trained and equipped to locate, identify, and mark drop/landing zones, provide limited weather observations, install and operate navigational aids and air traffic control communications necessary to guide aircraft to drop/landing zones, and to control air traffic at these zones.

<u>Combat Search and Rescue</u>: Combat search and rescue (CSAR) is a specialized task performed by rescue forces to effect the expeditious recovery of distressed personnel from a hostile environment during wartime or contingency operations.

<u>Combined Operation:</u> An operation conducted by forces of two or more allied nations acting together for the accomplishment of a single mission.

<u>Command and Control:</u> The exercise of authority and direction by a properly designated commander over assigned forces in the accomplishment of the mission. Command and control functions are performed through an arrangement of personnel, equipment, communications, facilities, and procedures employed by a commander in planning, directing, coordinating, and controlling forces and operations in the accomplishment of the mission (JCS Pub 1).

<u>Compact Laser Designator (CLD):</u> The compact laser designator is a target marking device with a rangefinder. The man-portable target marker will be used by a ground operator for target handoff to laser guided ordnance and laser tracker equipped aircraft. The CLD is a Class IV neodymium yttrium aluminum garnet (ND:YAG) laser. It weighs 16 lbs. and has a range from 50-1,000 meters. The primary power source is a lithium battery, although rechargeable nickel cadmium batteries are available for training.

GLOSSARY OF NAVAL SPECIAL WARFARE TERMS

Communications: A method or means of conveying information of any kind from one person or place to another (JCS Pub 1).

Compartmentation: Establishment and management of an intelligence organization so that information about the personnel, organization, or activities of one component is made available to any other component only to the extent required for the performance of assigned duties.

Compromise: The known or suspected exposure of clandestine personnel, installations or other assets, or of classified information or material, to an unauthorized person.

Counter-Guerrilla Warfare: Operations and activities conducted by armed forces, paramilitary forces, or non-military agencies against guerrillas.

Counterinsurgency: Those military, paramilitary, political, economic, psychological. and civic actions taken by a government to defeat insurgency (JCS Pub 1).

Counter-intelligence: Those activities which are concerned with identifying and counteracting the threat to security posed by hostile intelligence services or organizations or by individuals engaged in espionage, sabotage, or subversion (JCS Pub 1).

Counterterrorism: Offensive measures taken to prevent, deter, and respond to terrorism. Also called CT (Approved definition by JCS Pub 1).

Cover: Protective guise used by a person, organization or installation to prevent identification with clandestine activities.

Covert Operations: Operations which are so planned and executed as to conceal the identity of, or permit plausible denial by the sponsor under the provisions of Executive Order 12036. They differ from clandestine operations in that emphasis is placed on concealment of identity of the sponsor rather than on concealment

APPENDIX F.2

of the operation.

Deception: Those measures designated to mislead the enemy by manipulation, distortion, or falsification of evidence to induce him to react in a manner prejudicial to his interests.

Denial Operation: An operation designed to prevent or hinder enemy occupation of, or benefit from, areas or objects having tactical or strategic value.

Direct Action Mission (DAM): A specified military or paramilitary operation involving a commando style raid into a hostile or denied area. DAM's are usually conducted covertly or clandestinely by SPECOPS forces in order to rescue, strike, reconnoiter, or destroy a target behind enemy lines.

Diversion: The act of drawing the attention and forces of an enemy from the point of the principal operation; this can be an attack, alarm, or feint which diverts attention.

Drop Altitude: Altitude of an aircraft in feet above the ground at the time of a parachute drop.

Drop Zone (DZ): A specified area upon which airborne troops, equipment, or supplies are air dropped.

Electronic Intelligence (ELINT): The intelligence information product resulting from the collection and processing, for subsequent intelligence purposes, of foreign noncommunications electromagnetic radiations emanating from other than atomic detonations or radioactive sources.

Encipher: To convert plain text into unintelligible form by means of a cipher system.

Encode: 1. That section of a code book in which the plain text equivalents of the code groups are in alphabetical, numerical, or other systematic order. 2. To convert plain text into unintelligible form by means of a code system.

GLOSSARY OF NAVAL SPECIAL WARFARE TERMS

Encrypt: To convert plain text into unintelligible form by means of a crypto system.

Espionage: Actions directed toward the acquisition of information through clandestine operations.

Evader: Any person who has become isolated in hostile or unfriendly territory who eludes capture.

Evasion and Escape (E&E): The procedures and operations whereby military personnel and other selected individuals are enabled to emerge from an enemy-held or hostile area to areas under friendly control.

Evasion and Escape Net: The organization within enemy held or hostile areas that operates to receive, move, and exfiltrate military personnel or selected individuals to friendly control.

Evasion and Escape Route: A course of travel, preplanned or not, which an escapee or evader uses in his attempt to depart enemy territory in order to return to friendly lines.

Forward Operating Base (FOB): In unconventional warfare, a base usually located in friendly territory or afloat which is established to extend command and control or communications or to provide support for training and tactical operations. Facilities are usually temporary and may include an airfield or an unimproved airstrip. The FOB may be the location of JUWTF component headquarters or smaller unit which is supported by a main operating base.

Foreign Internal Defense: Participation by civilian and military agencies of a government in any of the action programs taken by another government to free and protect its society from subversion, lawlessness, and insurgency (JCS Pub 1).

APPENDIX F.2

Guerrilla Warfare: Military and paramilitary operations conducted in enemy held or hostile territory by irregular, predominantly indigenous forces (JCS Pub 1).

Harassment: An incident in which the primary objective is to disrupt the activities of a unit, installation, or ship rather than to inflict serious casualties or damage.

Human Intelligence (Humint): A category of intelligence derived from information collected and provided by human sources.

Insurgency: An organized movement aimed at the overthrow of a constituted government through use of subversion and armed conflict (JCS Pub 1).

Infiltration: 1. The movement through or into an area or territory occupied by either friendly or enemy troops or organizations. The movement is made, either by small groups or by individuals, at extended or irregular intervals. When used in connection with the enemy, it infers that contact is avoided. 2. In intelligence usage, placing an agent or other person in a target area in hostile territory. Usually involves crossing a frontier or other guarded line. Methods of infiltration are: black (clandestine); grey (through legal crossing point but under false documentation); white (legal).

Interdiction: Preventing or hindering by any means, enemy use of an area or route.

Intelligence: The product from the collection, processing, integration, analysis, evaluation and interpretation of available information concerning foreign countries or areas (JCS Pub 1).

Logistics: The science of planning and carrying out the movement and maintenance of forces. It incorporates supply and services, maintenance, transportation, ammunition, construction, and medical services (modified JCS Pub 1).

GLOSSARY OF NAVAL SPECIAL WARFARE TERMS

Low-Intensity Conflict: A limited politico-military struggle to achieve political, social, economic, or psychological objectives. It is often protracted and ranges from diplomatic, economic, and psychosocial pressures through terrorism and insurgency. Low-intensity conflict is generally confined to a geographic area and is often characterized by constraints on the weaponry, tactics and level of violence. Also called LIC (approved definition for JCS Pub 1).

Marker: A visual or electronic aid used to mark a designated point.

Marking Panel: A sheet of material displayed by ground troops for visual signaling to friendly aircraft.

Meaconing: A system of receiving radio beacon signals and rebroadcasting them on the same frequency to confuse navigation. The meaconing stations cause inaccurate bearings to be obtained by aircraft or ground stations.

Military Assistance Advisory Group: A joint service group, normally under the military command of a commander of a unified command and representing the Secretary of Defense, which primarily administers the US military assistance planning and programming in the host country. Also called MAAG (JCS Pub 1).

Military Civic Action: The use of preponderantly indigenous military forces on projects useful to the local population at all levels in such fields as education, training, public works, agriculture, transportation, communications health, sanitation, and others contributing to economic and social development, which would also serve to improve the standing of the military forces with the population. (US forces may at times advise or engage in military civic actions in overseas areas). (JCS Pub 1)

Net, Chain, Cell System: Patterns of clandestine organization, especially for operational purposes. Net is the broadest of the three; it usually involves (a) a succession of echelons and (b) such functional specialists as may be required to accomplish its mission. When it consists largely or entirely of non-staff employees, it may

APPENDIX F.2

be called an agent net. Chain focuses attention upon the first of these elements; it is commonly defined as a series of agents and informants who receive instructions from and pass information to a principal agent by means of cutouts and couriers. Cell system emphasizes a variant of the first element of the net; its distinctive feature is the grouping of personnel into small units that are relatively isolated and self-contained. In the interest of maximum security for the organization as a whole, each cell has contact with the rest of the organization only through an agent of the organization and a single member of the cell. Others in the cell do not know the agent, and nobody in the cell knows the identities or activities of other cells.

Net Authentication: An authentication procedure by which a net control station authenticates itself and all other stations in the new system systematically establishing their validity.

Overt Operation: The collection of intelligence openly, within concealment. Operations which are planned and executed without attempting to conceal the operation or identity of the sponsoring power.

Paramilitary Forces: Forces or groups which are distinct from the regular armed forces of any country, but resembling them in organization, equipment, training, or mission (JCS Pub 1).

Peacetime Contingency Operations: Politically sensitive military operations normally characterized by the short term rapid projection or employment of forces in conditions short of conventional war (e.g., strike, raid, rescue, recovery, demonstration, show of force, unconventional warfare and intelligence operations). (TRADOC Pam 525-44).

Propaganda: Any form of communication in support of national objectives designed to influence the opinions, emotions, attitudes, or behavior of any group in order to benefit the sponsor, either directly or indirectly (JCS Pub 1).

GLOSSARY OF NAVAL SPECIAL WARFARE TERMS

Radar Beacon: A receiver-transmitter combination which sends out a coded signal when triggered by the proper type of pulse enabling determination of range and bearing information by the interrogating station or aircraft.

Raid: An operation, usually small-scale, involving a swift penetration of hostile territory to secure information, confuse the enemy, or destroy his installations. It ends with a planned withdrawal upon completion of the assigned mission.

Recovery Site: An area within or outside a SAFE (E&E) area from which an evader/escapee can be evacuated. The area is selected for its accessibility by ground, sea, or airborne recovery personnel.

Sabotage: An act with an intent to injure, interfere with, or obstruct the national defense of a country by willfully injuring or destroying, or attempting to injure or destroy, any national defense or war material, premises, or utilities to include human and natural resources.

SAFE Area: A designated area in hostile territory which offers the evader or escapee a reasonable chance of avoiding capture and of surviving until he can be evacuated.

Search and Rescue: The use of aircraft, surface craft, submarines, specialized rescue teams and equipment to search for and rescue personnel in distress on land or at sea.

Sensitive: Requiring special protection from disclosure which could cause embarrassment, compromise, or threat to the security of the sponsoring power. May be applied to an agency, installation, person, position, document, material, or activity.

Sensitive Area: Specific location which has become a center of activity of intelligence interest.

APPENDIX F.2

Signal Panel: Strip of cloth used in sending code signals between ground and aircraft in flight.

Special Activities: Means activities conducted abroad in support of national foreign policy cojectives which are designed to further official United States programs and policies abroad and which are planned and executed so that the role of the United States government is not apparent or acknowledged publicly, and functions in support of such activities, but not including diplomatic activity or the collection and production of intelligence or related support functions.

Special (or Project) Equipment: Equipment not authorized in standard equipment publications but determined as essential in connection with a contemplated operation, function, or mission.

Special Forces Operational Base (SFOB): In unconventional warfare, a provisional organization which is established within a friendly area by elements of a Special Forces group to provide command, administration, training, logistical support, and intelligence for operational Special Forces detachments and such other forces as may be placed under this operational control. (Note: CINCPAC adds "The SFOB also provides logistical support for indigenous UW forces sponsored by those detachments. The Commander, SFOB will normally be the Army component commander of the JUWTF if only one SFOB is utilized.").

Special Operations: Operations conducted by specially trained, equipped and organized DOD forces against strategic or tactical targets in pursuit of national military, political, economic, or psychological objectives. These operations may be conducted during periods of peace or hostilities. They may support conventional operations or they may be prosecuted independently when the use of conventional forces is either inappropriate or infeasible.

Strategic Intelligence: Intelligence that is required for the formation of policy and military plans at national and international levels.

GLOSSARY OF NAVAL SPECIAL WARFARE TERMS

Strategic intelligence and tactical intelligence differ primarily in level of application but may also vary in terms of scope and detail (JCS Pub 1).

Tactical Intelligence: Intelligence which is required for the planning and conduct of tactical operations. Tactical intelligence and strategic intelligence differ primarily in level of application but may also vary in terms of scope and detail (JCS Pub 1).

Target: 1. A geographical area, complex, or installation planned for capture or destruction by military forces. 2. In intelligence usage, a country, area, installation, agency, or person against which intelligence operations are directed.

Target Acquisition: The detection, identification, and location of a target in sufficient detail to permit the effective employment of weapons.

Target Folders: The folders containing target intelligence and related materials prepared for planning and executing action against a specific target.

Terrorism: The unlawful use or threatened use of force or violence against individuals or property to coerce or intimidate governments or societies, often to achieve political, religious, or ideological objectives.

Theater: The geographical area outside continental United States for which a commander of a unified or specified command has been assigned military responsibility.

Transponder: A transmitter-receiver capable of accepting the electronic challenge of an interrogator and automatically transmitting an appropriate reply.

Unconventional Warfare: A broad spectrum of military and paramilitary operations conducted in enemy, enemy held, enemy controlled, or politically sensitive territory. Unconventional warfare includes, but is not limited to, the interrelated fields of guerrilla warfare, evasion and escape, subversion, sabotage, and other operations of a low visibility, cover, or clandestine nature. These interrelated aspects of unconventional warfare may be prosecuted singly or collectively by predominantly indigenous personnel, usually supported and directed in various degrees by (an) external source(s) during all conditions of war or peace.